World on Fire

World on Fire

Humans, Animals, and the Future of the Planet

MARK ROWLANDS

OXFORD
UNIVERSITY PRESS

Oxford University Press is a department of the University of Oxford. It furthers
the University's objective of excellence in research, scholarship, and education
by publishing worldwide. Oxford is a registered trade mark of Oxford University
Press in the UK and certain other countries.

Published in the United States of America by Oxford University Press
198 Madison Avenue, New York, NY 10016, United States of America.

Library of Congress Cataloging-in-Publication Data
Names: Rowlands, Mark, author.
Title: World on fire : humans, animals, and the future of the planet / Mark Rowlands.
Description: New York, NY, United States of America : Oxford University Press, 2021. |
Includes bibliographical references and index.
Identifiers: LCCN 2020058283 (print) | LCCN 2020058284 (ebook) |
ISBN 9780197541890 (hb) | ISBN 9780197541913 (epub)
Subjects: LCSH: Food of animal origin—Environmental aspects. |
Animal culture—Environmental aspects. | Climate change mitigation. |
Environmental health. | Deforestation.
Classification: LCC TD899.F585 R65 2021 (print) | LCC TD899.F585 (ebook) |
DDC 179/.3—dc23 LC record available at https://lccn.loc.gov/2020058283
LC ebook record available at https://lccn.loc.gov/2020058284]

DOI: 10.1093/oso/9780197541890.001.0001

Contents

Acknowledgments

Many of the ideas contained in this book first saw the bright light of day, or at least the gloomy twilight of my mind, in a course on environmental ethics that I regularly inflict on students at the University of Miami. I'd like to thank those students, in various iterations of this course, for their forbearance and longanimity, while I rambled on about EROIs, and carbon dioxide equivalences, extinction rates and fiscal devaluation in the late Roman Empire. Painful it must have been for you, I truly understand. But let me assure you that you were all essential to the birth of this book; midwives to it in the best Socratic sense.

If my teaching played an important role in the genesis of this book, then the absence of it played just as important a role in its completion. I spent the whole of 2019 utterly bereft of teaching responsibilities. Yes, an entire, glorious calendar year—the first time this has ever happened in my protracted professional career. The University of Miami's Center for Humanities was kind enough to furnish me with a fellowship, which took care of the spring teaching. A regular sabbatical then took care of the autumn. My thanks to all responsible parties.

I am grateful to Peter Ohlin, my editor at Oxford University Press, for his unfailing encouragement, with this and earlier projects. My thanks also to the Press's anonymous reviewers who made some very helpful suggestions.

Finally, as usual, my greatest debt is to my family: to my wife, Emma, and my sons, Brenin and Macsen. The latter flit in and out of this book, the butt of my clumsy but, I assure you, affectionate jokes. We are all starting to feel the beginnings of it. But it is their generation—Z, Alpha, and those who come after—that will bear the brunt of this monstrous befouling of the planetary bed; a ghastly mess perpetrated by predecessors not a lot unlike me, whether Millennial, Xennial, Xer, Boomer, Silent, Greatest, Interbellum, Lost, and so on and so forth back to the dawn of our species.

Sorry about that, boys! No, really. The least I can do, in the circumstances, is dedicate this book to you. But—let's face it, you're probably eating a chicken as I type these words—I don't think you are going to like it.

World on Fire

1

Animals, Energy, Land

An Overview

The Animals of Chernobyl

The scales fell from my eyes one day shortly after I had watched the HBO mini-series *Chernobyl*. Doubts have been expressed about its veracity, but, accurate or not, the series was certainly *gripping*. At least, I was gripped. Sufficiently gripped, in fact, to conduct a cursory Internet investigation of the Chernobyl exclusion zone of today, with particular reference to its flora and fauna. Things seem to be going rather well.[1] Within the roughly 2600 square kilometer (a little over 1000 square mile) exclusion zone—a rough circle with a radius of 30 kilometers—we seem to find a wildlife paradise. Bears, bison, boars, badgers, deer, eagles, and lots and lots of wolves, all formerly driven out by the area's human occupants, have now returned, living and repopulating, seemingly inconvenienced by the elevated levels of Uranium 235. Appearances can be deceiving, of course. Perhaps chronic or congenital health problems occasioned by generations lived inside the zone will eventually emerge. But, at present, the animal residents of the Chernobyl exclusion zone are certainly doing a very good impression of thriving. Indeed, they seem to be doing so well that we humans even decided to relocate an endangered species of wild horse—the Przewalski horse—into the zone, because we thought the horses might do a little better there than other areas where there was, you know, us. Thus far, it seems we were right—their numbers have increased spectacularly.[2]

Chernobyl is the site of the single worst episodic environmental disaster ever concocted by human hand. I label it "episodic" to distinguish it from

---

[1] See, for example, "Animals Rule Chernobyl Three Decades after Nuclear Disaster," *National Geographic*, April 18, 2016, https://www.nationalgeographic.com/news/2016/04/060418-chernobyl-wildlife-thirty-year-anniversary-science/.

[2] See "The Mystery of Chernobyl's Wild Horses," *The Conversation*, April 28, 2020 (updated May 4, 2020), https://theconversation.com/the-mystery-of-chernobyls-wild-horses-137270.

World on Fire. Mark Rowlands, Oxford University Press. © Oxford University Press 2021.
DOI: 10.1093/oso/9780197541890.003.0001

long-term problems—climate change, mass extinction, the sort of things this book is going to talk about in fact. Judged as a single episode, they don't seem to come any worse than Chernobyl. The conclusion seems inescapable. The animals have voted—voted with their feet, in fact. And as far as they are concerned, even the worst environmental disaster ever concocted by humans is better than the humans themselves. Ouch!

The aim of this book is to explore the place of animals in the various types of environmental malaise we face today. The animals in question divide into two broad types: *wild* and *domestic*. The types of environmental malaise divide into three: *climate change, mass extinction,* and *pestilence.* What connects these two categories, I shall argue, is a massive project that went hand in hand with the birth and development of human civilization: a *gigantic biomass reallocation program.* A biomass calculation for a given species is obtained by multiplying the mass of an average member of the species with the estimated number of individuals in that species. When we do this, as did Bar-On, Phillips, and Milo quite recently, we see a huge swing from wild to domestic biomass.[3] Today 96% of all mammalian biomass consists in humans and domestic farm animals—that is, the animals we have created so that we may eat either them or their products. That does seem rather excessive. Around 70% of all avian biomass consists in domestic fowl—equally stunning when you remember that human biomass is not factored into this percentage. The result of human civilization has been an astounding reallocation of biomass: we took it from wild animals and reassigned it to ourselves and to the animals we like to eat, or whose products we like to eat. The biomass of wild mammals has declined nearly 85% since our Neolithic forebears discovered farming. We did that.

The other category is made up of the three main environmental problems we face today. First, there is the problem of *climate change.* The climate is changing in ways that will profoundly affect the planet and its ability to support life in the sort of way to which we have become accustomed. The expression "global warming" has fallen out of favor—for largely spurious reasons—but we all know that the "change" the climate is undergoing is in a concerted and conspicuous warming direction. Indeed, because our current pestilential circumstances have taken up all of our attention, it is difficult to remember that back in January 2020, not insignificant swathes

---

[3] Y. Bar-On, R. Phillips, and R. Milo, "The Biomass Distribution on Earth," *Proceedings of the National Academy of Sciences* 115 (2018): 6506–6511. https://doi.org/10.1073/pnas.1711842115.

of the planet were on fire. There is, second, a problem of *mass extinction*: species of fauna and flora are currently dying off at a rate between several hundred and several thousand times the normal background rate of extinction. Some have claimed that the earth is entering into a *sixth great extinction*; this one, unlike the others, brought about by human activity. Third, there is, of course, the problem of *pestilence*. I did vacillate a little over this label. I initially favored *plague*. If I'd written this book in the 1300s, that might have worked. But, today, the label plague is, officially if not colloquially, restricted to diseases—bubonic, pneumonic, and septicemic—caused by the bacterium *Yersinia pestis*. So I went with *pestilence* (even though Y. pestis obviously derives from this). Suitably four horseman-y, I think. And, while even a few short months ago this label would have required some explanation, today, any such provision would, I assume, be otiose.

This book explores the extent to which the extraordinary biomass reallocation program brought about by human civilization is connected with these environmental problems we now face. The connections between the two, I shall argue, run surprisingly deep. As a result, substantial—perhaps startling—progress can be made on the problems of climate change, extinction, and pestilence if we are willing to revisit our biomass reallocation predilections. That is, we have to stop farming animals for their flesh and their products. If we are serious about addressing our environmental problems, biomass that is now allocated to domestic animals—cattle, pigs, sheep, chickens, and many others—should be reassigned to wild animals and, crucially, the land remade into a suitable home for them. That is the central contention of this book.

I shall—unless I'm feeling unusually sanguine in penning the closing stanzas of this book—stop short of the claim that this reassignation of biomass will put an end, once and for all, to the problems of climate change, extinction, and plague. But I shall argue that rethinking the biomass distribution on the planet will take us a surprisingly long way in the attempt to solve each of these problems. Not quite *sufficient* to solve them, perhaps, but, nevertheless, allowing us to travel a surprisingly significant portion of the distance to sufficiency. Moreover, it is almost certainly *necessary*. If we do not reverse the allocation of biomass, slant it heavily back in favor of wild rather than domesticated animals, it is likely that we will never really come to grips with these three major environmental threats. Necessary, then, and a significant portion of the way to sufficient, to address our environmental challenges. The distribution of biomass among the earth's animals is, I shall try to convince you, as important as that.

It's Complicated! Reasoning under Conditions of Profound Ignorance

Sometimes, it's not easy being an active, engaged citizen. This need not be through any deficit of intent or desire but, rather, because of the woeful quality of the information at one's disposal. Environmental issues provide very good examples. Much of what we are told about the environment is false or misleading. It has to be. Ocean levels will rise 1 meter by 2100. Or 2 meters. Or 60 meters. It can't be all of these. Nuclear fission has an energy returned on energy invested (EROI)—the ratio of energy it yields against the amount of energy we need to put into the building, running, maintenance, and fueling of nuclear power stations—of more than 85. Or less than 1. Species are currently disappearing at a rate of 100 times the normal background rate of extinction. Or 10,000 times. A kilogram of beef protein requires 6 kilograms of vegetable protein to build it. Or 25. Or more than 54 kilograms. Animal agriculture produces less than 5% of the world's greenhouse gas (GHG) emissions. Or 14.5%. Or 51%. Many—perhaps most—estimates in a given domain will turn out to be not just be wrong but massively so.[4] Much of what we are told about the environment should, it seems, be treated with a pinch of salt.

How does one think and reason about the environment under these conditions of profound uncertainty? Indeed, characterizing these conditions as ones of *uncertainty* is, perhaps, overly optimistic in many cases. On a common way of thinking about it, uncertainty about an issue presupposes a reasonably well-defined and/or accepted answer, perhaps framed within a certain range, with a certain degree of residual doubt about the probability we can assign to this answer. Sometimes, the concept of uncertainty can be legitimately applied. For example, different models of future global temperature trajectories forecast increases of 3°C–5°C by the end of the century.[5] This is on a *business-as-usual* scenario—that is, if we keep on doing what we

---

[4] I shall be citing the work from which these figures, and the ones below, derive in later chapters. In the interests of flow—i.e., not clogging up the pages with footnotes when I shall do precisely that later in the book—I shall rescind from footnotes at this juncture. Given their role here in merely illustrating procedures for reasoning under conditions of ignorance, you can, for now, simply treat them as (actual or possible) "for instances," if you like.

[5] I am thinking, in particular, of RCP 6 and RCP 8.5. RCP stands for Representative Concentration Pathway. See Chapter 3 for discussion.

have been doing. The variation between 3°C and 5°C is the result of different initial assumptions, some of which are more plausible than others. There are quite a few different ways of keeping on doing what we have been doing—and therefore, more than a few different business-as-usual scenarios. In this case, we have a range of possibilities, and probabilities can be assigned to each point in the range, depending on the initial assumptions we make. This, then, can be legitimately described as a case of uncertainty. In other cases, however, this description is dubiously applicable. A range of 1 to >85 for the EROI of nuclear fission, for example, is not really an answer range as such, but an array of wildly divergent answers. The same is true of estimates of species numbers—vital to know if, for example, you want to know the rate at which species are becoming extinct—which range from less than 5 million to over 100 million. In such circumstances, we are not so much dealing with conditions of profound uncertainty but, rather, those of profound ignorance. The position is especially grim for the nonspecialist who may not be in a position to reliably assess the often very technical work required to produce such estimates.

There are certain rules I have followed to help mitigate the effects of these epistemically malign circumstances. They are, individually, fallible, but collectively, I have reasonable confidence they will help in the required epistemic mitigation process—certainly in the long run if not in every case.[6]

First, in so far as this is possible, *restrict oneself to peer-reviewed* work.[7] More precisely, restrict oneself to peer-reviewed work for any claim on which one is going to rest any substantial weight—that is, if the claim is important to the argument one wishes to develop.[8] There is, as we all know, an awful lot of crap out there. The restriction to peer-reviewed work ensures that at some sort of quality control has gone into assessing the work whose estimates you are going to use in your attempts to reason about the environment. It may still be rubbish, but the chances of this have been reduced considerably. This, at best, provides partial epistemic mitigation only. It is

[6] I am sure the irony will not be lost on anyone here. In order to mitigate the effects of profound uncertainty and/or ignorance in reasoning about environmental issues, I am adopting rules in which I have "reasonable confidence." That is, I am adopting rules to which I am willing to assign a certain probability of being effective in order to mitigate the effects of claims that can only be assigned a certain probability of being true. There is no reason to stop there. How "sure" am I that the irony will not be lost on anyone? That would be a probabilistic matter, too. In this game, it is probabilities—or "confidence"—all the way down, and also all the way up.

[7] Or—which pretty much amounts to the same thing—to claims presented by government and professional bodies (e.g., NASA, NOAA, EPA, IPCC, IPBES, World Bank, etc.) that can reasonably be understood as reporting peer-reviewed research.

[8] So, for example, animals in Chernobyl—just introductory filler, not essential to any argument I am going to develop. I can use *National Geographic* for that.

not as if the peer-reviewed work on our three environmental problems all concurs on the answers to the important questions and, by sieving out the chaff of the non-peer-reviewed cyber-sphere, you can now press forward confidently with solid, unequivocal information you can use as premises in our argument. On the contrary, much of the challenge of (and, indeed, interest in) assessing claims about the environment derives from profound disagreements in the peer-reviewed work published in respectable journals.

Second, *outliers are not to be trusted.* A defeasible rule—totally: not only are outliers sometimes correct, what counts as an outlier often changes over time. For example, what counts as an outlier in estimates of sea-level rise by 2100 has been trending upward in recent years, as have estimates about global temperature increase by 2100 on a business-as-usual scenario. Nevertheless, the distrust of outliers is a useful rule of thumb in many circumstances, especially when combined with Rule 1 that militates against the non-peer-reviewed. An outlier that hasn't been peer-reviewed? One can only treat this claim with extreme distrust.

Rule 2a, or perhaps 1a: I don't think it is sufficiently important, or, indeed, separate from Rules 1 and 2, to merit elevation into a separate rule of its own, but here it is, for what it's worth: *motivation should not be ignored.* It is *never* a bad idea to ask yourself, "Who is sponsoring this work?" Nuclear fission has an EROI of >85 you say? But your work is sponsored by the World Nuclear Association? Hmmm. No, it's <1 you say? But you are lifelong, extremely vocal, critic of nuclear power. Similarly, hmmm. If, with respect to a piece of work, you can add *non-peer-reviewed outlier* to this concern about motivation, there is no obvious reason to take the claims made in this work very seriously. That is not to say there are no reasons at all—no *obvious* reason is not the same as no reason. But it does mean that a lot of work will have to be done by the purveyors of these reasons before one should begin to take them seriously. It is comparatively rare for such work to be successful.

Third, and most important so far: *don't make life easy for yourself.* In fact, make life hard for yourself. You are, let us suppose, faced with a range of estimates for a given phenomenon. Some of those estimates suit your dialectical purposes very well indeed. Others are far less convenient. Then, go with the less convenient. For example, suppose estimates of the carbon sequestration capacities of a given type of tree—the amount of carbon dioxide trees of this type can capture in a year—range from 48 to 88 pounds. And suppose also that it would be absolutely *peachy* for your dialectical purposes if trees of this type, on average, sequestered 88 pounds of carbon dioxide a year. Then, I am afraid you are obligated to assume the 48-pound average. Maybe you don't have to be a complete masochist. But certainly don't assume anything more than 50 pounds. The skill of any dialectician worth his or her salt is to get one's arguments to succeed *even in the most inopportune of circumstances*. You, in effect, are saying: look, help yourself to whatever premises you want (within reason), I, or rather my arguments—let's pretend this is not gladiatorial—are still going to win.

Fourth, my favorite and therefore the one I choose to regard as most important: *try to see the underlying picture*. Data, information, estimates, and so on and so forth: if they are correct, accurate, or veridical, they record the way the world is. Far more important than this is the way the world has to be. Sometimes the most important thing one can do to organize one's thinking is to identify a picture. This picture allows you to frame, contextualize, and so understand data, information, estimates, and the like and so put them in the logical and conceptual places they are supposed to be. In the opening section, I began to provide such a picture when I organized the natural world in terms of *biomass*. But mass—biomass or otherwise— we know is equivalent to *energy*. Another, very useful—indeed, in the context of climate change in particular, perhaps the most illuminating— picture of the environment can be painted using a palette provided by the concept of energy.

The Continuity of Food and Fuel

If the calculations of Bar-On, Phillips, and Milo are correct, a rather striking 96% of the mammalian biomass on earth today consists in humans or domestic animals, the latter raised to be eaten by humans or to provide products that humans will eat. On top of that, there is the 70% of avian biomass that comprises domestic birds, raised for the same reason. We know, courtesy of Einstein, and long before him an ancient pre-Socratic philosopher named Heraclitus—that mass is equivalent to energy.[9] Thus, we might reframe these biomass statistics in terms of the

---

[9] Okay, so Heraclitus didn't say precisely that. But he said certain things very suggestive of it. I'll talk a little about Heraclitus, and what he did say, in Chapter 4.

concept of energy. This, I shall argue throughout the course of this book, is enormously useful. Energy is the concept we need to discern the deep connections between our use of animals, climate change, mass extinction, and pestilence.

First, to state the obvious, climate change is the result of certain energy choices we have made. Historically, we have chosen to use stored sources of solar energy—embodied in trees, coal, oil, and natural gas—to solve our energy needs. This is not unreasonable, because these fossil fuels are excellent—that is, highly *concentrated*—sources of energy. The expression "energy needs" is entirely appropriate. We *need* energy because we are complex structures. A combination of the First and Second Laws of Thermodynamics—(1) energy can be neither created nor destroyed but merely converted from one form into another, plus (2) the disorder of an energetically isolated system always tends to increase—yields an unequivocal conclusion: if any complex structure (such as you or I) is to continue to exist, it needs to take in energy. Without this, it breaks down—we die. What should not be overlooked, however, is that humans, and other biological organisms, are not the only complex structures. Societies are also complex structures and, as such, also subject to the first two laws of thermodynamics. We acquire the energy we need for the continuation of ourselves and our societies from two sources: the fuels we burn and the food we eat. Therefore, the particular distribution of energy across the planet, reflected in biomass distribution, is the result of choices we have made both for ourselves and, by proxy, the societies we inhabit: choices about how to acquire the energy we need to keep ourselves, and the societies in which we are embedded, in existence.

To keep ourselves warm, or cool, and to keep our industries and transport systems thriving, we have chosen to burn fossil fuels. To keep ourselves fed, we have chosen to breed, raise, and kill cattle, pigs, chickens, and other domestic animals on an industrial scale. Both are energy choices and, as such, entirely continuous. This idea of the continuity of food and fuel is central to the arguments I shall develop in this book. What I mean by calling them *continuous* is that they are *versions of the same sort of process performed for the same purpose*. The *process* is one of energy acquisition and utilization. The *purpose* is the continuation of certain complex structures—ourselves and the societies we inhabit. The energy from food and that from fuel are both parts of the same overall story: the story of how certain complex structures acquire the energy they need to continue in existence. As such, they can be assessed together and, when it is illuminating to do so, balanced against each other.

These two energy choices—fossil fuels and animals—are ones with distinct downsides. The burning of fossil fuels has resulted in the climate crisis that we now face. But the eating of animals, farmed on industrial scales, is also an energy choice and, in its way, just as disastrous. It is a disastrous energy choice in that for every joule of energy we obtain from, for example, the flesh of cattle, we have to invest anywhere between 6 and 54 joules of energy to obtain it. These kinds of upside-down energy choices would never have been permitted when we were countenancing which fuels to employ to heat our homes and power our industries. If we had to invest 20 times as much as energy into retrieving coal as we obtained through burning it, then the burning of coal would never have caught on. Disastrous energy choices will always, eventually, have consequences. It is these we are witnessing today in the form of climate change, mass extinction, and plague.

Food, Extinction, and Pestilence

There is an energy choice which is bringing about significant change in our climate, which is likely to have serious impacts on human well-being and that of the biosphere more generally. There is also another energy choice that, at least prima facie, seems to be ridiculously inefficient in the sense that far more energy has to be into building these stores of energy than can ever be harvested from them. The two choices can't really be separated: they are both attempts made by complex structures to keep themselves, and the complex structures in which they are embedded, in existence. Suppose, now, you were told that you had to give up one of these choices. Of course, there is no need to think this is really an either/or decision. It may well be that we will have to give up both. But suppose—for now just suppose—we could get away with giving up only one of them. Which one would you sacrifice? More importantly: which one *should* you sacrifice?

In the context of climate change, there has been much more focus on the first choice. We need to stop burning fossil fuels. Stop driving big cars. Use public transport. Perhaps even, God forbid, walk. Then there is *Flygskam*: we should be ashamed of stepping onto planes, these metal birds of astronomical carbon footprints. I don't necessarily disagree with any of this, although I do hope that some of the more extreme forms of this sentiment do not turn out to be required. But this first type of choice is not the subject of this book. The focus of this book is on how far we can push the second choice: giving up our disastrously inefficient food-based energy choices. The more effective is the second choice in curbing climate change the less need will there be for the first choice—at least in its more extreme incarnations.

The first choice—austerity with regard to fuel consumption—is focused squarely on the problem of climate change. The second choice—essentially, abstemiousness with regard to one's food choices—is an interesting one, in part, because it promises to help address not just this but also the other two environmental problems: mass extinction and pestilence. Consider, first, the problem of mass extinction. The disastrously inefficient energy choice that consists in the raising, killing, and subsequent eating, of animals on an industrial scale is not something that can be achieved in a vacuum. Certain changes need to be brought about in order to pursue this choice. In particular, certain changes have to be wrought in the land. What is primarily forestland must be converted into what is primarily pasture land. And the more humans there are—we are currently closing in on eight billion—and the more accustomed we are to eating meat, then the more land will require conversion. This is what's known as a *change in land use*. Changes in land use are a critical component of the pursuit of the vast biomass reallocation project that we know as animal agriculture. The question that we should then consider is obvious: what is the connection, if any, between changes in land use and species extinction?

The United Nations recently convened a body to look at the issue of extinction. This body is known as IPBES: the Intergovernmental Science-Policy Platform on Biodiversity and Ecosystem Services, which, admittedly, is a bit of a mouthful. Their report, released in 2019, argued that change in land use is, by some distance in fact, the most significant cause of species extinction today.[10] Forests, per hectare, carry far more species than pasture land. The change from forest to farmland almost always entails a reduction in biodiversity, undermining previously intact ecosystems. Furthermore, while changes in land use are partially driven by human urban expansion, this is not their major cause. Rather, the primary cause

---

[10] IPBES, "The Global Assessment Report on Biodiversity and Ecosystem Services," https://ipbes.net/global-assessment.

of change in land use is agricultural expansion. Finally, as the report also makes clear, the vast majority of agricultural expansion today is pastoral rather than arable: most agricultural expansion today is pursued in order to produce meat. In contrast with agricultural expansion, climate change, according to IPBES, ranks as only the third most important driver of species extinction. If our concern, therefore, was with species extinction, as well as climate change, we might be well advised to focus our attention on the second sort of energy choice—the one that involves converting vegetable matter into animal matter and then eating the latter. The argument for this will be developed in Chapters 8 and 9.

This focus also yields another benefit. As I type this sentence (August 2020), COVID-19, caused by the SARS-CoV-2 virus, is killing hundreds of thousands of people around the planet. These are official death figures, from cases where death has resulted subsequent to a positive test for the virus. As preliminary estimations of excess death count in several countries indicate, the real death toll is likely to be much higher, and it may already run into the millions. While the exact provenance of COVID-19, and its precise pathway into the human population, is still disputed, there is little doubt that COVID-19, like other coronaviruses, including SARS-CoV-1 and MERS-CoV, ultimately has an animal origin. In this, COVID-19 is entirely typical. According to the World Health Organization (WHO), well over half of all human diseases are zoonotic or have a zoonotic origin. That is, they are caused by pathogens that were originally carried by animals that subsequently jumped the species barrier into us. Moreover, the vast majority, and perhaps all, newly emerging diseases are zoonotic.[11] Temperate diseases overwhelmingly tend to derive from domestic animals: cattle, pigs, chickens, etc.[12] In general, since domestication is as old as human civilization, many of these diseases have been around for quite a while and we have adapted to them. They inconvenience us, sometimes fatally so, but only rarely occupy the front page of the newspapers. That is, until a new one comes along. Remember mad cow disease? Tropical diseases, however, tend to originate in wild animals. Since we are less accustomed to them, they do tend to be more dramatic and so get a lot more journalistic attention. The HIV virus, for example, which, according

---

[11] World Health Organization, "Neglected Zoonotic Diseases," https://www.who.int/neglected_diseases/diseases/zoonoses/en/. As we shall see in Chapter 10, other estimates elevate these percentages quite significantly.

[12] N. Wolfe, C. Dunavan, and J. Diamond, "Origins of Major Human Infectious Diseases," *Nature* 447 (2007): 279–283. https://www.nature.com/articles/nature05775.

to the WHO has killed over 30 million people worldwide since the outbreak began in 1981,[13] was acquired from chimpanzees and other primates, where it took the form of SIV, simian-immunodeficiency virus. The human predilection for bushmeat likely provided the opportunity for this virus to jump species.

With regard to human diseases, then, the overall picture looks like this. Most of them we acquire from animals. For us to acquire them from animals, there must be appropriate opportunities for the relevant pathogen to jump species. The likelihood of a pathogen jumping species is a function of, among other things, the frequency of encounters between the original host species and the potential new host—namely, us. When we farm animals, we, obviously, encounter them frequently. When we trespass onto lands ancestrally occupied by certain wild species—either because we want to clear their land for pasture, or because we plan on turning them into bushmeat—we similarly encounter animals at elevated frequencies. Moreover, the result of transforming land ancestrally occupied by various species into pastureland will often be the displacement of these species, which may then wander on to regions occupied by us, thus, again, upping the frequency of encounters. Therefore, I shall argue, to reduce the likelihood of further zoonotic diseases, the most significant course of action we can take is to abandon our practice of eating animals. This argument will be developed in Chapters 10 and 11.

Austerity with regard to fuel choices is a response to climate change. But climate change is not the only critical environmental problem we face. There are also the problems of mass extinction and plague. Changes in our food choices seem to be far better placed to positively impact the latter two problems. For many, however, the acid test will be how well changes in our food choices are able to blunt the problem of climate change.

Food and Climate Change

Amending our food choices can, I shall argue, also significantly mitigate the problem of climate change. This is so for two reasons. The first has to do with the continuity of fuel and food outlined earlier. In Chapters 4 and 5, I am

---

[13] World Health Organization, https://www.who.int/hiv/data/en/.

going to develop the following line of argument. Ultimately, if we are serious about addressing the problem of climate change, there is no alternative to renewable energy. However, as things stand, renewable energy—involving extremely diffuse energy sources—stands, at best, on the cusp of viability. It may be capable of meeting our energy needs, or it may not. It really could go either way. What renewable energy technology needs is a little *breathing space*: time to develop into the mature, effective technology it promises to be. That time is best provided, I shall argue, by the precise targeting of a particular portion of the energy supply train that keeps us humans and our societies in existence. This targeting consists in abandoning the ridiculously inefficient practice of converting energy stored in plants to energy stored in animals, and then consuming the latter. This is a luxury we can, in our present circumstances, no longer afford. Through abandoning the practice of raising and killing animals for food on industrial scales, we thereby take out—according to well-accepted, mid-range estimates of the United Nations—roughly 14.5% of current GHG emissions. That, in itself, is not a solution to the problem of climate change, but it will buy us the time needed for renewable technology to develop, hopefully, into mature technology with improved levels of efficiency in production and storage. This argument is developed in Chapter 6.

This breathing space—the buying of time—is not, however, the only benefit of abandoning animal agriculture. Indeed, it is not even the greatest benefit. The result of the disastrous energy choice that we have made—raising animals for food, where the energy we get out of this arrangement is by several orders of magnitude less than the energy we have to put in—is, as is often the case with ill-advised energy transactions, is that we have to make available vast quantities of land to raise the animals we have chosen to eat. This is the most significant factor driving the problem of mass extinction. At the same time, the increased frequency of encounters with animals—wild and domestic—through farming them and destroying their ancestral homes is the most significant factor driving the recurrence of zoonotic diseases. Also, however, and somewhat crucially, it also robs us of an enormous opportunity to offset our monstrous size 16½ carbon footprint.[14]

Imagine if we ceased our continual expansion into the ancestral homes of animals, ended out relentless conversion of forest into pastureland. More

---

[14] The significance of this rather specific number will become apparent later, in Chapter 6 especially.

than that: imagine if we reversed it. The second consequence is that vast amounts of land will become newly available. Most of the land currently utilized by humans is used either for grazing the animals we eat or for growing crops—typically soy—to feed the animals we eat. If we abandon our habit of eating meat, much of this land becomes available for other uses. Admittedly, some of this land will be taken up by the extra crops we have to grow to feed ourselves rather than the animals we formerly ate. However, the energy conversion ratio of plants to animals entails that the amount of land we would have to devote to growing these extra crops for human consumption would be vastly less than the amount of land we save by no longer eating animals. This land, no longer needed to pursue our disastrous energy exchange with animals, could be returned to its ancestral state.

Much of this ancestral state, though not all, will have been forest. After the last ice age, trees grew pretty much wherever they were able. Not all land was suitable for forest. For example, the Pampas of South America and the Great Plains of North America were too dry to support trees, except in their numerous riparian corridors. But after the last ice age, and before Neolithic and Bronze Age farmers began their remorseless project of clearing the forest, parceling it into ever smaller and smaller packages, much of the world was carpeted in thick forest. Suppose we returned this land to its former state. When the newly forested land reaches maturity, it will provide a carbon sink that can significantly offset our climate emissions. I shall argue, in Chapter 7, that the carbon footprint of the most energy-hungry society in history can be significantly lessened, indeed substantially erased, if we give up our practice of raising animals for food and return the land that we no longer need to the way it was before we started doing this.

You Broke It, You Fix It! Or, Perhaps, Share and Share Alike

The central message of this book is that we should stop eating animals and the products of animals. This message is directed, primarily, at the peoples of the most developed portions of the globe—Europe and North America in particular. It is we—I speak as a European who has lived much of his adult life in North America—who are most responsible for the environmental threats we all now face, and we who have benefited most from the exploitation of the environment that created these threats. You could look at these facts in a nakedly retributive way: You created this mess; therefore, you get us out of it. Or you could look at it more in terms of the concept of fairness: You have already more than used up your share of carbon-based resources, so please leave some for everyone else. The climate crisis is a product of the industrialization undergone, in the first instance, by the West. It is hypocritical, therefore, to criticize other regions of the world for adding to the problem by doing exactly the same thing we did not so long ago. Much criticism has been directed toward Brazil in recent years, especially since the election of Jair Bolsonaro, for deforestation of the Amazon rainforest. Perhaps Bolsonaro is not doing enough to stop it. Or perhaps he is actively encouraging it. But, you have to admit, he makes a good point: "Few countries have the moral authority to talk about deforestation with Brazil. . . . I would like to give a message to the beloved Angela Merkel. Take your dough and reforest Germany, okay? It's much more needed there than here."[15] Or, as Luis Inácio da Silva, a prior Brazilian president of rather different political persuasion, put it a few years earlier: "The wealthy countries are very smart, approving protocols, holding big speeches on the need to avoid deforestation. But they already deforested everything."[16] We did. Admittedly, two wrongs don't make a right. But it is very difficult to point a finger at someone committing one of those wrongs, when you committed the other one not so long ago.

The central contention of this book is that we all need to stop eating animals. But it is not for me to tell poor Congolese villagers, for example, that they have to stop eating bushmeat. For them, there may be no choice. But I know that for us, in the West, and increasingly in other parts of the developing world, there is a choice. Given our role in engendering the three environmental threats we all face today—through our empire of fossil fuels and our conspicuous consumption—it is incumbent on us, first and foremost, to put our house in order. We can do this, fundamentally, by abandoning a habit that has been around as long as we have: the habit of eating animals and their products. Don't misunderstand me. I think it would be a good thing if everyone were to abandon this habit. And I hope that, one day, everyone will. But this exhortation is made, in the first instance, to everyone who is in a position to endorse and adopt it. To those of us who can: we must stop eating animals and their products.

[15] "Jair Bolsonaro to Merkel: Reforest Germany, Not Amazon," *Deutsche Welle*, August 15, 2019, https://www.dw.com/en/jair-bolsonaro-to-merkel-reforest-germany-not-amazon/a-50032213.
[16] "Brazil's Lula Blasts Rich Nations on Climate," *Reuters*, February 6, 2007, https://www.reuters.com.

Taking Dreams Seriously

The role played by our farming practices in driving major environmental threats has hardly gone unnoticed. However, it is rare to find anyone willing to take seriously the idea that we should simply abandon the practice of eating animals. When the United Nations' Intergovernmental Panel on Climate Change (IPCC) churns out special report after special report, identifying, in very useful detail, the malign effects on the climate of our farming practices—in particular, our *pastoral* farming practices—the issue of simply abandoning our carnivorous habits never comes up. Instead, we must change the way we farm animals, amend our farming practices, perhaps eat less meat. But eating no meat at all is—clumsy pun obviously intended—not on the table. Similarly, when IPBES put out their long-awaited 2019 special report on extinction, the punchline was that farming animals for their food and their products was, by some distance, the primary cause of species extinction. Farming practices needed to be amended. Our meat-eating habits needed to be dialed back. But the idea that we should simply stop eating meat was, apparently, never on the menu.

To reduce the risk of any further bad puns, let's just consider a question: Doesn't that seem strange? Is the idea of no longer eating animals and their products so outrageous that it can't even be broached—even though it seems to be a logical course of action to adopt in the face of our three gravest environmental threats? It can be done quickly, too—and alacrity is crucial in addressing these threats. The average lifespan of a broiler chicken is 6 weeks. That of an intensively raised pig is 5 months. The main alternative is a form of *austerity*, broadly construed. We must dramatically scale back our other carbon emissions—from the industrial, power, and transport sectors. Transport, in particular, has taken a lot of flak in recent years, especially aviation. It is doubtful that austerity of this sort would even address the issues of extinction and pestilence. As we shall see, climate change is only the third most important driver of extinction, and its role as a driver of pestilence is also dwarfed by animal agriculture. But even if we restrict ourselves to climate change, questions have to be asked about the relative desirability of this kind of austerity.

Which, for example, is more desirable? Amending our eating habits? Or drastically changing our transport options? In the end, life is all about needs and dreams, and I suspect transport is far more thoroughly entwined with both of these than meat. My children need to go to school. They might disagree, but the last 6 months of lockdown have convinced me that they do. Miami was built around the car. In this, it is like many American cities: cars and cities over here tended to grow up together; they coadapted to each other. Integrated transport systems weren't part of the plan and, quite frankly, are proving a little difficult to implement now—the massive construction required would interfere with all the cars! And, therefore, a car is, more or less, a necessity. I need to drive my children to school. I need to drive to work, as does my wife. These are all needs. Besides needs, there are dreams. One of my sons dreams of being a goalkeeper for a major European football team, and so, after school, he requires me to ferry him to his training sessions, which, lately, can often be 50 miles away. We could get a bus, but it would take over 24 hours to do the round trip. I could say to my son: Okay, no school or footy for you—climate change, mate. But, then, his life would be a much poorer one. A life where needs are not met and dreams cannot be indulged is a poorer life. His dream is an improbable one, of course, but that doesn't matter. What does matter is this: if he does realize his dream of being the last line of defense and first line of attack, for Barcelona, or Real Madrid, or PSG, or Manchester United—I can guarantee you, with 100% certainty, this dream will not specify what he gets to eat after the match. Dreams rarely involve food. Unless you are starving, of course.

Planes are great, because they take us away from the crappy places we live to somewhere better, if only for a while. My mother never set foot on a plane in her life. She didn't even venture abroad—by way of a car ferry—until she was nearly 50 years old. Her life, in my opinion, was smaller because of it. I mean no disrespect. I took my first flight, and my first foreign travel, when I was 14 years old—a cheap package skiing holiday organized by my school in the Swiss Alps. (How cheap? We flew from Luton Airport, mate.) The months leading up to it went by in an agony of anticipation. I had dreamed of this for so long. I dreamed of the snow, the slopes, of new culture unknown to me, and of exotic 14-year-old girls. I did not dream of what I was going to eat when I got there.

Dreams are, admittedly, a privilege. But remember for whom I said this book is written. That would be us, in the West, or the developed world more generally. Poverty is still a tenacious problem, of course. But if someone's dreams were to revolve around eating meat, this would simply be an indication of just how badly his or her life had gone to this point. That is one of the great evils of grinding poverty: it makes your dreams small. But the antidote to privilege is to spread it around as much as possible. Everyone should be able to have dreams, and, more pertinently, they should be big ones. Expansive ones. It may be that, in the end, *Flygskam*—and other serious amendments to our transport options more generally—is our only option. But let's not pretend this would be anything other than a tragedy. In the face of impending tragedy, it is always wise to explore other options. To avoid my *cancelation*, let me be clear that I am *not* claiming that we don't have to rethink our attitude to planes, and fossil fuel emissions more generally. But that is not what this book is about. Rather, in this book, I shall try to convince you that abandoning our ancient habit of eating animals is probably the best thing we can do in face of the three potential environmental calamities that beckon—these three threats understood collectively rather than individually. Abandoning this habit is a course of action that we should take very seriously because it is easier to implement, more effective when implemented, and, ironically, more *palatable* than the alternatives.

2

Good Morning, Atlantis!

Why I Don't Like Mornings

I must admit: I have become very attached to Miami. I have lived here for
13 years and counting—that's longer than I have lived anywhere in my adult
life. My children were born here, and it feels like home. As much as anywhere
has ever felt like home. Perhaps when you come from the United Kingdom
and end up in Miami, you're not going to miss the old country too much. Not
the horizontal rain and hail, or the icy damp that slowly eats its way into your
bones. Only the mountains, really. Or the hills. Any sort of gradient would be
nice. But, apart from that, I have few complaints. It's easy to love the weather,
the ocean, and the beaches, of course. But even the sky-high grocery prices, the
mercurial caprices of the drivers, ranging from boy racers to heavily medicated
nonagenarians, and the astonishing rudeness of large swathes of the populace: I
regard all of these with great fondness. I don't pick and mix Miami. I embrace
it all. I have watched with something akin to paternal pride as the adolescent
city in which I arrived in January 2007 has slowly evolved into something more
mature, more substantial. Miami now has a recognizable downtown. It didn't
in 2007. There are even the nascent stirrings of an integrated public transport
system, although I'm not going to hold my breath on that. Much of this development has been driven by a gargantuan influx of foreign capital, especially from
South America, but also Europe, east and west. And sometimes, in the evening
when the sun grows long, and a margarita or two has been given a good home,
I can almost convince myself of something: Miami is now too big to fail. It must
be too big to fail. But when I wake up in the morning, I know this is not true.

I hate the mornings. The mornings are when I remember that Miami is,
in essence, an affront to nature. In the mornings, I remember that building a
city on a coast that rarely peeks its head more than 6 feet above sea level in the
middle of hurricane alley was probably not one of humankind's better ideas.
I don't know if it'll happen in my lifetime,[1] but, sometime soon—maybe not

---

[1] For me at least, one of the more disconcerting facets of COVID-19 is its potential to cancel my old
age—a cancelation virus for a cancelation culture. Ironic. COVID-19 aside, as the recently canceled

World on Fire. Mark Rowlands, Oxford University Press. © Oxford University Press 2021.
DOI: 10.1093/oso/9780197541890.003.0002

tomorrow, maybe not the day after, but soon—Miami is going under. At least, it will if we don't seriously and rapidly amend our ways.

That Sinking Feeling

With Miami's seemingly manifest destiny a doleful, hovering presence at the back of my mind, my attention often meanders toward practical matters. In particular, how long will it be, exactly? Not how long until Miami goes under but, rather, how long until people cotton on to the fact that it will? Like a recently rediscovered packet of ramen noodles that has been sitting at the back of the pantry for too long, the remnant of a hurricane supply kit of years gone by, Miami's expiration date is coming around with noticeable rapidity. How long will it be, I wonder, before this date becomes general knowledge and my house, accordingly, becomes worthless? There is a cheerful little website I like to visit from time to time, when I manage to rustle up a quiet moment or two. Here it is. Or, if you are reading the print version of this book—how very quaint—the web address can be found in this footnote.[2] If you are also a resident of a coastal city, it will tell you all about the future prospects of your home, too.

I'll spare you the details, since if I start talking about land east of the Turnpike and west of the Turnpike, and north of the Dolphin and south of the Dolphin, few who live outside south Florida will have any idea what I am talking about. If the Climate Central website is to be believed, the general contours of the submersion of south Florida will look something like this. After 1 foot of sea-level rise, around a third of the Everglades will disappear. Parts of Miami Beach will also start to look distinctly soggy. After 2 feet of sea-level rise, around one-half of what used to be the Everglades will be absorbed into the Gulf of Mexico. New areas of swampland will start to appear, especially in southern Miami-Dade and western Broward counties. After 3 feet of sea-level rise, three-quarters of what used to be the Everglades will lie beneath the Gulf of Mexico. The "New Everglades," which started to appear in southern Miami-Dade after 2 feet of sea-level

comedian, Louis C.K., once remarked, you never really know whether you're in the middle of a healthy life or at the end of an unhealthier alternative.

[2] https://ss2.climatecentral.org/#11/25.7617/-80.1918?show=satellite&projections=0-RCP85-SLR &level=1&unit=feet&pois=hide.

rise, will now be absorbed by the rapidly expanding Florida Straits. After 4 feet of sea-level rise, almost all of the former Everglades will disappear. Miami Beach will be essentially no more. The same is true, with the possible exception of certain parts of North Key Largo, of the Keys. I'll miss The Keys. Farewell *Margaritaville*. Adios Jimmy Johnson's *Big Chill*. Hasta la vista *Sparky's Magic Landing*.

But never mind! My house in Palmetto Bay is still there! People told me I was a fool to worry about this, but I knew that buying a house 5 feet above sea level—that's positively vertiginous territory in these parts, you know—would pay off in the end. Indeed, if my eyes don't deceive me, the Climate Central map now has it as oceanfront property! I might even still have a job: the "U"—the University of Miami—will still be there. Admittedly, I will need a boat to get to work, but I've wanted one of them for years. It's not a frivolous extravagance now, is it darling? And I can even tie it up outside the house! It's true that the "U" might find it difficult to attract nonlocal students: Miami International Airport will now be at least partially submerged. Through 5, 6, 7, and 8 feet of sea-level rise, the boat will be all I have left. What was once South Florida progressively transforms into the "New Keys," islands that become smaller and smaller with each new inch of sea-level rise. My oceanfront residence will disappear at around 6 to 7 feet of rise, the canal onto which it backs having merged with the Atlantic. After 10 feet of sea-level rise, a few tiny slivers of land aside, that will be pretty much all she wrote. South Florida will sleep with the fishes.

The Rock Disaster Scale

Details aside, the general contours of this sort of disappearance will be replicated in thousands of coastal cities around the world. Princeton ecologist Benjamin Strauss, in a recent publication in the *Proceedings of the National Academy of Sciences*, agrees, including Miami on a list of 414 cities across the United States that will be beneath sea level by 2100.[3] Of course, being beneath sea level doesn't necessarily mean being underwater. Just ask New Orleans. But, as New Orleans can also attest, it does mean that you are rather disposed

---

[3] B. Strauss, S. Kulp, and A. Levermann, "Carbon Choices Determine US Cities Committed to Futures below Sea Level," *Proceedings of the National Academy of Sciences USA* 112, no. 44 (2015): 13508–13513.

to periods of an underwater variety should an unfortunate event or two transpire. Three-quarters of the world's megacities are on the coast, and roughly 40% of the world's population lives on or within 60 miles of the coast.[4] Not all coastal areas are as relentlessly flat as South Florida, of course. But more than a few are, and in one form or another, the rise of the oceans is going to be a problem for everyone. So, the question that should be on everyone's lips—or, at least, on the lips of nearly 40% of the world's population—is obvious: just how much sea-level rise should we expect in the foreseeable future, say, the next 50–100 years?

The short answer is: *it depends*. The long answer is: *it depends on a variety of things*. Sea-level rise is being driven by global temperature rise: temperature increase averaged out over the whole planet. And global temperature rise is being driven by the increased concentration of greenhouse gases (GHGs) in our atmosphere. These gases include carbon dioxide, methane, nitrous oxide, and chlorofluorocarbons (CFCs). I'll talk about all of this in the next chapter. So how much the seas will rise depends on what we do to the atmosphere in the coming decades.

"Business as usual" is the expression used to mean carrying on pretty much as we are now. Actually, there are many ways we can carry on pretty much as we are now. So there are more than a few business-as-usual scenarios. When people talk about *the* business-as-usual scenario, what they are usually talking about is the average of all the possible business-as-usual models. It's a bit like hurricane models. In August and September, whenever a tropical disturbance appears in the Atlantic, we South Floridians all start keeping an apprehensive eye on the *models*. These are models of the projected path of the disturbance and its expected strength—whether it will become a tropical storm or a hurricane—and if the latter whether it will be a run-of-the-mill hurricane or a major hurricane (i.e., category 3 or above). Both path and strength are difficult to predict. Each model makes different initial assumptions, and these result in different projected paths and intensities. The National Hurricane Center (NHC) helpfully provides a map here,[5] and, whenever a tropical disturbance is in the offing, on this map will be a cone. The cone designates anywhere the hurricane/storm/depression might be, and any point within the cone is as likely as any other. The center of the cone, however, is the predicted path calculated

---

[4] Michael Harris, *A Future for Planning: Taking Responsibility for Twenty-First Century Challenges* (New York: Routledge, 2019), 21.
[5] National Hurricane Center and Central Pacific Hurricane Center: nhc.noaa.gov.

from averaging out all the different models. This changes as data are acquired and the models are updated. In a lot of cases, the models will eventually all converge and then we can all either (a) get the hell out of Dodge, (b) hunker down, or (c) feel slightly guilty at our relief that some other poor bastards are going to get it this time.

Business as usual is, then, the average of various different business-as-usual models. This average—common or garden—business as usual is then indexed to a time. The year 2100 is a common one. The year 2050 is becoming more common because of a growing suspicion that climate change might be happening rather more quickly than we thought. But let's go with 2100. If we adopt *business as usual*, what will sea levels be like in 2100? The answer we get—like the projected path and intensities of hurricanes—depends on a number of assumptions. Until very recently, standard estimates of sea-level rise by 2100, on an average business-as-usual scenario, were almost comforting—ranging from 0.52 to 0.98 meters by the year 2100.[6] That's roughly 1 feet 8 inches to 3 feet 2 inches. That's not too bad, one might think. I get to keep my (perhaps, newly oceanfront) house, and maybe even my job—if I live to 138 that is. Too bad about the Everglades, of course, and the Keys will probably be a little dicey. But, as environmental cataclysms go, this one seems very much on the tame side.

There is, as far as I know, no unified scale for environmental disasters. There are scales for different types of disasters. Nuclear disasters, for example, are rated on a scale of 1 to 7. To date only Chernobyl and Fukushima Daiichi have been rated 7s. But there is no scale for environmental disasters in general. That's a disgrace, and it changes here. There is a clear, easy, and entirely scientifically objective way to measure the seriousness of an environmental disaster. You simply ask yourself: would Dwayne Johnson be willing star in a movie about this? The strength of your conviction that he would map neatly onto the severity of the environmental catastrophe in question. A massive earthquake, as in San Andreas: yes. A giant albino gorilla, and an equally large wolf and alligator *rampaging* their way across the North American continent: you bet Dwayne is going to be in that. But 3 feet of sea-level rise? That would elicit what is known in the business as a *hard pass* from Dwayne's people. His agent wouldn't even

---

[6] J. Church et al., "Sea Level Change," in *Climate Change 2013: The Physical Science Basis. Contribution of Working Group I to the Fifth Assessment Report of the Intergovernmental Panel on Climate Change*, ed. T. Stocker et al. (New York: Cambridge University Press, 2013), 1137–1217. https://www.ipcc.ch/site/assets/uploads/2018/02/WG1AR5_Chapter13_FINAL.pdf.

bother sending him the script. We might, for reasons entirely obvious to anyone who knows who Dwayne Johnson is, call this scale of measurement the *Rock* scale. Three feet of sea level would barely register on the Rock scale. However, if you are in the business of making disaster movies, and want the Rock to star in them, all is not yet lost.

It's Not like a Bathtub

Just as with the hurricane models, predictions of sea-level rise depend on what initial assumptions we make. And I think there is a general consensus emerging that some of the assumptions made by the standard estimates are, at best, questionable and perhaps unjustifiable. If we replace these assumptions with more realistic ones, sea-level rise might, after all, be of interest to Dwayne's people.

The first assumption is not one made by standard estimates as such, but it does tend to feature in how people understand the significance of these estimates. The assumption is that the world's oceans are something like a bathtub, with the same water level everywhere. In fact, however, this is not so. The surface of the ocean is not flat but lumpy. I'm not talking about waves and local disturbances like that. Rather, the level of the world's oceans is higher in some places than in others. The result is that a 36-inch global rise in sea level might translate into, say, 30 inches in some places and 42 inches in others. There are several reasons why my adopted hometown is likely to be on the higher end of this.

First, large land masses, such as the Greenland Ice Sheet and the West Antarctic Ice Sheet, exert a gravitational attraction on the surrounding water, essentially pulling the water toward them, resulting in lower sea levels in other places, such as the Tropic of Cancer, for example, or, for that matter, 25.62°N, 80.32°W. In other words, if the great northern and southern ice sheets melt, the resulting reduction in their gravitational pull will see water flow from the higher latitudes—where sea levels will actually drop—toward the lower ones. Sea-level rise will be higher in places such as South Florida whose levels were formerly kept lower by the gravitational pull of the high-latitude ice masses.[7]

---

[7] According to NASA, "The loss of ice mass causes a resulting drop in sea level close to the ice sheets, while at a greater distance, the tropics face ice-related sea-level rise of about 20 percent more

Second, the warmer water gets, the more it expands.[8] One implication of this is that sea levels will tend to rise more in places with warmer water. In itself, this is an already prevailing condition. That is, it already applies to regions such as South Florida, and so shouldn't, on its own, cause sea level to rise more in this region than it already has. But suppose global temperatures increase. The water gets warmer than it is, and so expands more. In this way, we should expect sea levels to rise most in places with the warmest water.

Third, the centripetal force created by rotation of the earth on its axis will push the melting water from the poles toward the equator and form a kind of bulge there. As in the case of heat-driven water expansion, this is an already prevailing condition. However, to the extent that ocean water has moved from the higher latitudes to the lower ones, and to the extent this water has further expanded through warming, we should expect the effect of this centripetal force on the ocean level to be accentuated further.[9]

In short, a 3-foot global sea-level rise is likely to translate into something more than 3 feet for Miami. NASA estimates that the tropics face sea-level rises of around 20% higher than the global mean.[10] If correct, a global mean sea-level rise of 36 inches is likely to translate into 42–43 inches in places like Miami. Nothing in this process of estimation lends itself to any degree of certainty or exactitude. But it doesn't matter because I am just getting warmed up—no pun intended, this time. Standard estimates tell us that we should not expect more than a 3 foot global sea-level rise by 2100. But, recently, some there have emerged some rather worrying considerations, which, if they pan out, would mean that should expect quite a bit more than this.

---

than the global mean." See https://sealevel.nasa.gov/understanding-sea-level/regional-sea-level/water-mass-change. They also supply a nice little computer simulation that allows you to see local changes in sea level over the past few decades as the result of the melting of ice sheets.

[8] See https://sealevel.nasa.gov/understanding-sea-level/global-sea-level/thermal-expansion. Also https://sealevel.nasa.gov/understanding-sea-level/global-sea-level/thermal-expansion. For some additional reasons for thinking that the Atlantic will be particularly affected by this phenomenon, see J. Krasting, J. Dunne, R. Stoufer, and R. Halberg, "Enhanced Atlantic Sea-Level Rise Relative to the Pacific under High Carbon Emission Rates," *Nature Geoscience* 9 (2016): 2010–2014.

[9] R. S. Nerem and J. Wahr, "Recent Changes in the Earth's Oblateness Driven by Greenland and Antarctic Ice Loss," *Geophysical Research Letters* 38, no. 13 (2011) L13501.

[10] For a useful study of the likely variations in sea-level rise occasioned by the loss of the West Antarctic Ice Sheet, see J. Mitrovica, N. Gomez, and P. Clark, "The Sea-Level Fingerprint of West Antarctic Collapse," *Science* 323, no. 5915 (2009): 753.

A Confidence Game

There is a crucial assumption that guides standard estimates of sea-level rise: even in a business-as-usual scenario, the melting of the West Antarctic Ice Sheet (including the Thwaites Glacier) would make only a minimal contribution—a few tenths of a meter at most—to sea-level rise. This is an assumption to which standard estimates ascribe "medium confidence." Climate science is a science of "confidence"—yes, a "confidence game" or "con" some, not me, would say—which, of course, comes in greater and lesser degrees. This is because it faces the unenviable task of modeling phenomena when the evidence it has amassed is incomplete, and where many of the most important variables required for the model to work are simply not known. Accordingly, climate scientists ascribe various degrees of "confidence" both to the models they have produced in these less-than-ideal circumstances and to the assumptions that they have made in developing these models.

The reason that scientists felt they could ascribe "medium confidence" to the assumption that the melting of the West Antarctic Ice Sheet and Thwaites Glacier would make only a comparatively small contribution to sea-level rise was because there was no known model of how ice sheets and glaciers melt that would license any more than this modest estimate. However, in a paper published in the journal *Nature* in 2016, Robert DeConto and David Pollard questioned this assumption.[11] Their argument came in two parts: the first focusing on historical evidence, and the second on a revised, and more comprehensive, model of how ice sheets melt.

I think it is becoming generally recognized that one of the problems with models of sea-level rise is that they do not pay enough attention to historical evidence. One of the merits of DeConto and Pollard's approach is that they address and rectify this deficiency. Their historical case is based on analysis of two periods. The first was the *Pliocene*—a period ranging from 5.3 to 2.6 million years ago—where atmospheric carbon levels were more or less the same as they are today: around 400 parts per million (ppm) during the Pliocene, whereas today they are 416.18 ppm.[12] The second period was the *Eemian*—an interglacial period from 130,000 to 115,000 years ago—where global temperatures were roughly at today's levels (slightly, but only slightly, higher). DeConto and Pollard pointed out that during the Pliocene, with

[11] R. M. DeConto and David Pollard, "Contribution of Antarctica to Past and Future Sea-Level Rise," *Nature* 531 (2016): 591–597.
[12] Data accessed on May 28, 2020. See https://www.co2.earth/.

similar levels of atmospheric carbon, sea levels were up to 30 feet higher than they are at present. During the Eemian, with global temperatures similar to today, sea levels were between 24 and 32 feet higher.

This is a softening-up argument. Its purpose is to create a *presumption* in favor of a higher estimate for sea-level rise than is countenanced in standard estimates. Given today's atmospheric carbon dioxide levels and global temperature profile, the historical record shows that far higher sea levels are what we should expect, all things being equal. Crucially, however, if sea levels were indeed this much higher in the Pliocene and Eemian periods—particularly the Eemian, which, like today, was an interglacial period—this could only have come from melting of Antarctic ice that is far greater than predicted in standard estimates. This brings us to the main part of DeConto and Pollard's argument.

Models of Melting: Marine Ice Shelf Instability and Marine Ice Cliff Instability

Jakobshavn (Inuit: *Sermeq Kujalleq*) is Greenland's largest glacier and the main outlet for the Greenland Ice Sheet. It is also melting, and this melting is driven by twin features that are central to DeConto and Pollard's account: *marine ice shelf instability* (MISI) and *marine ice cliff instability* (MICI).[13] Where a glacier meets the ocean, we typically find an *ice shelf*—a floating promontory that grows out from the front of the glacier and helps stabilize it by protecting it from the warmer ocean water. Jakobshavn, however, no longer has an ice shelf. Any ice shelf faces two potential adversaries. One is the warmer ocean water, which can erode it from below. The other is the warmer air above, which can cause the surface to melt. This surface meltwater, often augmented by rain, causes further problems, most notably *hydrofracture*. Surface meltwater collects in pools and then leaks into fissures. Eventually, these fissures open, and a section of the ice shelf breaks off. This is known as *calving*. Marine ice shelf instability, and subsequent calving, is a well-documented phenomenon and a perennial feature of the Arctic and Antarctic since the end of the last ice age. Calving itself will lead to little change in sea levels since the ice

---

[13] DeConto and Pollard did not focus on Greenland's Jakobshavn but, rather, on the Antarctic and Thwaites Glacier in particular. The direction my discussion has taken is perhaps driven by my enthusiasm for Northern Hemisphere glaciers, born during a visit to Snaefellsjokull, in Iceland, many years ago.

that breaks off is already either floating or has displaced seawater. However, the novel feature of DeConto and Pollard's account is the idea of MICI. Once the ice shelf has disappeared, larger ice cliffs will now be exposed to warmer seawater. Since the ground on which glaciers rest generally slopes downward, away from the ocean (that is, the glacier becomes thicker the further inland you go), this allows warmer ocean water to trickle under the bottom of the cliff—the portion below the waterline—slowly undermining the cliff. The exposed part of the cliff is also susceptible to the processes—surface melt-water accumulation, crevassing, and hydrofracture—that destroyed the ice shelf. The result is that the ice cliff collapses. This exposes the next part of the ice cliff (which will slowly be getting bigger as the glacier thickens, both from above and below, the further inland you go), and the same processes begin over again. Ice cliff collapse, since it involves much ice that is neither floating nor submerged, will have a significant effect on sea level.

Precisely how much of an effect is a matter of debate. DeConto and Pollard's original estimate was we should expect slightly more than an additional meter of sea-level rise over and above the 0.52–.98 meters countenanced in the standard estimates. That is, by 2100, we should expect sea levels to rise by up to 2 meters compared to current levels. As we have seen, the difference between 1 meter and 2 meters might be the difference between a severely compromised, Everglades-less South Florida and a South Florida that, effectively, no longer exists. Of course, sea levels won't stop rising just because we have reached the arbitrary date of 2100. DeConto and Pollard estimated an additional 13 meters of sea-level rise by 2500. These rises will not occur because of what is going on with Jakobshavn and the Greenland Ice Sheet more generally. Rather, the idea is that we can expect similar processes of MISI and MICI to operate in the Antarctic also, initially on the West Antarctic Ice Sheet (which is smaller and more vulnerable than the East Antarctic Ice Sheet) and in particular on the enormous Thwaites Glacier, which is several times bigger than Jakobshavn.

DeConto and Pollard's estimates have been contested in a recent article by Tamsin Edwards and colleagues, also published in the journal *Nature*.[14] Edwards's case is based on the controversial character of the MICI hypothesis and more sophisticated modeling that predicts a lower and slower rate of sea-level rise. This debate is a complex, technical one, and it depends, in

[14] Tamsin Edwards et al., "Revisiting Antarctic Ice Loss Due to Marine Ice-Cliff Instability," *Nature* 566 (2019): 58–64.

part, on how much confidence you like to place in models. In this context, an equally recent study by NASA is a little perturbing. *Operation Ice Bridge* is a NASA-led initiative involving satellites fitted with ice-penetrating radar. The result was the discovery of a massive cavity beneath the Thwaites Glacier—roughly two-thirds the area of Manhattan and 300 meters tall.[15] This strongly confirms the, unexpectedly rapid, disintegration of the glacier. Importantly, no model predicted this. It is not just the size of the cavity that is important but its alarming rate of growth. The cavity is big enough to have contained 14 billion tons of ice, and most of this ice has melted over the last 3 years.

Much of this debate is technical and can only be properly conducted by scientists. However, certain conclusions seem safe to draw. No one disputes that Jakobshavn is melting at a rapid rate. No one disagrees that Thwaites is disintegrating. While MICI has not been observed in the Antarctic, it has, reportedly, been observed in Jakobshavn.[16] All models agree that glacial melting will accelerate under current conditions. No one disagrees that, on a business-as-usual scenario, sea-level rises measured in meters rather than centimeters, feet rather than inches, are eventually going to occur. To the extent there is disagreement it concerns the timeframe for these changes. Some think the ice sheets and glaciers will collapse quite quickly. Others think this process will be more prolonged. The greatest source of uncertainty is due to the current rate of climate change. The earth regularly cycles between glacial (i.e., ice ages) and interglacial periods. But this cycling is gradual, taking place over tens of thousands of years. As we shall see, we are changing the climate—warming the earth—much more quickly than this. Current changes are occurring in decades—tens of years rather than tens of thousands of years.[17] This rate of change is historically unprecedented—completely unlike anything that has ever occurred before. This introduces some uncertainty, but it is doubtful that this uncertainty is going to be very comforting. An unprecedented rate of change may be *compatible* with the slow disintegration of ice sheets and glaciers. But it hardly *supports* this idea. In fact, it is not unreasonable to suspect the direction of support will run the other way.

[15] This is part of NASA's Operation Ice Bridge project, an "airborne campaign beginning in 2010 that studies connections between the polar regions and the global climate." See https://www.jpl.nasa.gov/news/news.php?feature=7322.

[16] J. Bassis. D. Benn, and B. Berg, "Anatomy of the Marine Ice Cliff Instability," paper presented at the Fall 2018 meeting of the American Geophysical Union.

[17] Admittedly, exits from glacial periods do tend to be relatively rapid. But, still, nowhere near as rapid as the warming the earth is currently undergoing.

Anyway, 2 meters is 2 meters, whether it occurs in 2100 or 2150. At some point, on a business-as-usual scenario, we can safely conclude that my house will sink beneath the sea. It will be bye-bye to the Miami-Fort Lauderdale metropolis. South Florida (like much most of its residents I assume) will relocate north, to begin somewhere in the vicinity of Boca Raton (although that itself will have moved a few miles inland). The Old Keys will have long gone, and what used to be Old South Florida has become the New Keys.

This may all sound very grim. But, in fact, we haven't even begun to do grim yet. There are two things to note about this issue of sea-level rise. First, there are some considerations that suggest that even the DeConto and Pollard estimate of 2 meters is unduly conservative.[18] We'll get to these in the next chapter. Second, the question of how much sea levels will rise—2 feet, 3 feet, 6 feet—is largely irrelevant to Miami and South Florida. The vast urban conurbation that spreads upward from Homestead, through Miami, Fort Lauderdale, and up to Hollywood is going to be gone—abandoned— long before it sinks beneath the waves. And it will all go out with a whimper rather than a bang.

A Superfund Here, a Superfund There

There's a part of me, a rather unpleasant part—a part of me that I really don't like very much at all—that feels strangely privileged. This is the part of me who feels he is living in antediluvian Atlantis. If I can just flog a few more decades out of this body, then I picture grandchildren sitting on my knee, asking me to tell them what it was like to live in Miami, the fabled lost city. Of course, there may well be another 400 or so fabled lost cities in the United States alone, rendering Miami just not that big of a deal. Still, I can do my best to big it up for the grandchildren. It would help immensely if Miami could go out in some dramatic sort of way. That's the way I picture the end of Atlantis—a vengeful ocean hurtling in, obliterating much, and submerging what remains. It might be good for my future narrative purposes if this could happen to Miami on some particularly apt day—Columbus Day being an obvious candidate. Columbus Day in Miami is a huge boat-based party. Immense flotillas of boats, captained and passengered by the very drunk and

---

[18] Indeed, if these reasons turn out to be accurate, even using the word "unduly" to describe how conservative is this estimate is probably also unduly conservative. *Catastrophically* conservative might be better.

the frequently naked, take over Biscayne Bay. One Columbus Day, perhaps, the Atlantic, tired of being urinated in and vomited on, rids itself of this city of degenerate, Dionysian excess. I was there, too, I'll tell them, in the boat I will surely have bought by then, but had the sort of lucky escape that even the youngest and most gullible of grandchildren will quickly find difficult to swallow. Except, you know, this is not how it's going to happen. Miami's sinking beneath the waves is actually the least of its worries. It will sink, prob-ably, but it's everything that happens before the sinking that's the problem. Miami's demise will be in no way as dramatic or glamorous as this. If the ocean did, one day, sweep in like this, the chances are there would be no one around to see it. Instead, the last days of Miami are likely to be damp, de-pressing, and, in all likelihood, rather disgusting.

Miami is a very thirsty city. This is not entirely surprising as, for much of the year, its residents have to deal with temperatures of 95°F and humidity levels of near 100%. That's a lot of water that needs replacing every day, and this water is obtained from the Biscayne Aquifer. The Biscayne Aquifer is not a good aquifer. Let me explain what a good aquifer looks like. The first thing to understand is that an aquifer is not an underground lake. An aquifer is made up of permeable rocks and other porous materials. Water is able to move through this saturated area. In the Biscayne Aquifer, for example, water generally moves in an east-southeasterly direction at a rate of around 2 feet per day. A good aquifer will be separated from groundwater by a layer of rel-atively impermeable rock, such as granite. Then, the aquifer is referred to as a *confined* aquifer. The drawback to confined aquifers is that they take a lot of time to refill naturally: decades if there is one layer of relatively imperme-able rock between the aquifer and the groundwater, and centuries if there are two such layers. But the advantage is that the water contained in a confined aquifer is very pure—much cleaner than, for example, the water in ground-level reservoirs. There are virtually no bacteria or other contaminants in con-fined aquifers as these have been filtered out on the long journey through the soil and rock.

The Biscayne Aquifer is not a confined aquifer. It can't be because there is no relatively impermeable rock, such as granite, in south Florida. Miami is built on the notoriously permeable limestone. So what is the difference between the Biscayne Aquifer and groundwater? The answer is: *none*. The Biscayne Aquifer—the only source of drinking water in south Florida—is, essentially, groundwater. I am not making this up. Here is the Miami-Dade official government website:

> The Biscayne Aquifer is located just below the surface of the land in South Florida. It is made out of porous rock with tiny cracks and holes. Water then seeps in and fills these tiny cracks and holes. This water is often referred to as groundwater or the water table, and provides virtually all of the water that is used by South Florida residents, visitors and businesses. . . . Because this drinking water supply is so close to the surface (barely a few feet down in most places), it is especially prone to contamination.[19]

The possible sources of contamination are many and varied. A *Superfund* site is one designated by the United States Environmental Protection Agency (EPA) as a location containing substances that are sufficiently hazardous to require long-term decontamination. Miami-Dade has had its fair share of these over the years, as has Broward, and several of these have contaminated the Biscayne Aquifer with such lovely substances as pesticides, metals (such as cadmium, iron, lead, zinc, and copper), and heavy metals (such as selenium, arsenic, and mercury).[20] A byproduct of limestone quarrying in Northwest Miami-Dade in particular is benzene, a well-known carcinogen, and this has also found its way into the Biscayne Aquifer. In 2005 one of the Northwest wells in Miami-Dade registered five times the federal limit for benzene.[21]

The concern is that these cases of water pollution are going to be significantly exacerbated by climate change. Since the 1960s, the amount of precipitation that falls during the heaviest storms has increased by roughly 7% in Miami-Dade. Seven percent can be the difference between a simple soaking and an outright flood. A 2014 EPA report warned that flooding from these more intense and frequent storms could push toxins from the contaminated soil of Superfund sites into the Biscayne Aquifer.[22] As you might recall, the Biscayne Aquifer is, in essence, groundwater. Therefore, it is not difficult to see how this could happen.

These sorts of threats are what we might think of as sporadic or intermittent. A Superfund site here, a Superfund site there. These threats could,

---

[19] https://www.miamidade.gov/water/biscayne-aquifer.asp.

[20] The EPA has a list of Florida Superfund sites here: https://www.epa.gov/superfund-redevelopment-initiative/superfund-sites-reuse-florida.

[21] See the testimony of Bill Brant, then director of Miami-Dade County Water and Sewer Department: http://media.miaminewtimes.com/2009148.0.pdf and http://media.miaminewtimes.com/2009136.0.pdf.

[22] "U.S. Environmental Protection Agency: Climate Change Adaptation Plan," publication number EPA 100-K-14-001, https://www.epa.gov/sites/production/files/2016-08/documents/epa-climate-change-adaptation-plan.pdf.

perhaps, be solved if the sites were properly cleaned up—if we assume that is technologically possible—or if limestone quarrying were halted, and so on. These intermittent threats are exacerbated by climate change. But it is climate change itself and the resulting sea-level rise that provide the most consistent threat to the Biscayne Aquifer.

Water, Water Everywhere

Miami, as I have mentioned, is built on limestone, and limestone is porous. That's why it's not possible to *build a wall* to keep the ocean out, even though wall building is popular in some quarters today. The ocean water would simply push its way in underneath us. Indeed, that's what it is doing now and has been doing for some time. Every year, as the sea level rises, saltwater pushes just a little further west, into the Biscayne Aquifer, threatening the region's drinking water.[23] A system of canals, higher than sea level, is used to control this intrusion. But the higher sea levels become, the higher the canals have to be, and then they lose their ability to control flooding. And we're back to the issue of sinking again. Moreover, saltwater can actually intrude into the aquifer from three different sources. First, it comes from the Atlantic Ocean and Biscayne Bay. Here it percolates underground, through the limestone rock, forming what is known as a *cone of ascension*. It's not really a cone, more a half parabola. Near the coast, the saltwater will intrude at all levels, from sea level to hundreds of meters below the ground. As the saltwater intrudes further inland, there is progressively less intrusion at shallower levels. The result is that as we reach the limits of inland intrusion, the saltwater is located only at the deeper levels of the aquifer. This is, ultimately, because saltwater is heavier than freshwater. However, there are two other sources of saltwater contamination of the aquifer: infiltration from tidal marshes and leakage from unprotected canals—ironically, some of the very canals that are used to control flooding and saltwater intrusion.[24] The intruding saltwater from these sources will eventually make its way down to the lower depths of the aquifer. But this is a gradual process, and at any given time saltwater may be

---

[23] See, for example, the 2014 United States Geological Survey report, "Origins and Delineation of Saltwater Intrusion in the Biscayne Aquifer and Changes in the Distribution of Saltwater in Miami-Dade County, Florida," https://www.nrc.gov/docs/ML1621/ML16216A235.pdf.

[24] USGS Report, "Origins and Delineation of Saltwater Intrusion in the Biscayne Aquifer and Changes in the Distribution of Saltwater in Miami-Dade County, Florida."

present at any depth in the aquifer (although not in the sort of concentrations you generally find at lower levels).

The Biscayne Aquifer was already susceptible to saltwater intrusion because of decades of pumping and draining water in the service of providing both drinking water and flood protection. But sea-level rises are predicted to increase the rate of intrusion at least twofold. In Miami-Dade, the intrusion of saltwater has almost reached the Hialeah Wellfield, one of the three main wellfields serving the city, and there is now almost total reliance on two larger wellfields to the west, the West Wellfield and the Northwest Wellfield. Broward has already lost some of its coastal wells to intrusion from the Atlantic and expects to lose over 40% of its coastal well capacity within the next 50 years.[25] The general strategy of dealing with this problem—establishing wells on the western edge of Miami-Dade and Broward Counties—does appear rather myopic. We should remember that with a mere 3 feet of sea-level rise, the Gulf of Mexico will be lapping at what will then be the new shores of western Miami-Dade and Broward. The long-term prospects of the western well strategy, therefore, do not look particularly promising.

And at the Other End . . .

Saltwater intrusion into aquifers is a problem for drinking water. But this intrusion also causes a problem at, in effect, the other end of the consumption chain.[26] Saltwater intrusion will also push up the groundwater level. The level of groundwater is crucial for any household—and, as we shall see, in Miami-Dade this amounts to a lot of households—that has a septic tank rather than being hooked up to the central sewer system. A septic tank works by collecting wastewater and solids into a large tub. The waste just sits there for a while, and gravity more or less efficiently does the work of separating out the solid from the liquid, pulling the latter out into what is known as the *septic drain field*, an area of soil beneath the tank. This liquid slowly filters through the soil, hopefully cleansing it of contaminants before it reaches the

---

[25] For a well-researched, popular account of the problems Miami faces, see Christopher Flavelle, "Miami Will Be Underwater Soon: Its Drinking Water Could Go First," *Bloomberg Businessweek*, August 29, 2018, https://www.bloomberg.com/news/features/2018-08-29/miami-s-other-water-problem.

[26] See Flavelle, "Miami Will Be Underwater Soon: Its Drinking Water Could Go First." Also, another popular and entertaining account, Mario Ariza, "Miami, Sea-Level Rise Is Coming for Your Poop," *The New Tropic*, June 25, 2017.

groundwater. However, this will only work if there is enough separation between the drain field and the groundwater. The State of Florida requires at least 2 feet of soil between the bottom of the drain field and the groundwater. However, during the wet season, and during king tides, this threshold is now regularly breached. When this happens, one or more things may occur. First, the liquid from the septic tank decides it has nowhere to go and so backs up into the tank, which then pours out into your yard. That's bad. Even worse, however, is the second possibility: the liquid decides it does have somewhere to go after all. The result is that untreated human waste passes into the groundwater. And, remember, the Biscayne Aquifer is an unconfined aquifer—the Biscayne Aquifer and the groundwater are essentially the same thing.[27] Thus, into the city's drinking supply go such lovely things as *E. coli* bacteria (which can cause diarrhea, vomiting, kidney failure, and death among other things), high levels of nitrates (which can cause, among other things, blue baby syndrome in infants), and pretty much every medication taken by every household of Miami-Dade—by my guess, the most highly medicated municipality in which I have ever resided—that is not hooked up to the central sewer system. In fact, one of the one of the ways in which researchers track septic-tank contamination is by tracking the levels of acetaminophen in the groundwater.

Investment and Return

It is not that there aren't any solutions to these problems. There are technological solutions that are available. The problem is that they are going to cost you. Investment and return: that is what the death of Miami will be all about. We shall see in later chapters that the ideas of investment and return are crucial drivers of environmental problems and their possible solutions. The notion of investment can cover many things—all of the many and variegated things one might potentially invest. In later chapters it is investment of energy and return on that investment that will take center stage. However, here, in the case of Miami, the version of investment is straightforward, mundane, financial investment.

---

[27] Recall the miamidade.gov website: "Because this drinking water supply is so close to the surface (*barely a few feet down in most places*), it is especially prone to contamination." Italics are mine.

Put yourself in the position of one of the nearly 100,000 households in Miami-Dade that have septic tanks rather than mains drainage. Your yard is covered in human waste. You can pay a company $250 for a pump-out, and probably do the same the next wet season, or king tide, and the one after that in perpetuity. Or you can pay $4000 to get the drain field fixed, at least for a while. What do you do? $4000 is a significant outlay, especially in a city that already has well-known affordability issues.[28] While there are pockets of extreme wealth in Miami-Dade, the median household income of the county is $44,000 plus change, and this is almost 25% lower than the countrywide average.

It would be much better for you in the long run, of course, if you could get your house connected to the central sewer system. But if you are balking at $4000, then the $40,000 to $50,000 it will cost to hook up an already built residence to the central sewer system is probably going to be beyond your budget. Moreover, if you are a homeowner with a drainage field problem, then you are also likely to have other problems, such as a susceptibility to frequent flooding during king tides, storms, and heavy rains. Throwing $40,000–$50,000 at a house such as this looks a lot like throwing good money after bad: a bad return on your investment. It would seem far more logical to pay $250 for the pump-out and put the house on the market.

Can't the County pay, you might wonder? Indeed, why shouldn't they pay? Arguably, they helped create the mess, during the massive explosion of development in southeast Florida, by permitting developers to adopt the cheaper option of installing septic tanks rather than connecting the houses they were building to the central sewer system. The reason the County won't, and probably can't, pay is that they simply can't afford it. As a result of these earlier decisions, Miami-Dade today has around 90,000 households with septic tanks. Even at the lower end estimate of $40,000 per household, the bill comes in at a whopping $3.6 billion. This amount of money is not going to be found without raising the price of water or raising taxes elsewhere. And, therefore, the relevant County commissioners are going to have to ask themselves some rather searching questions, such as the following: Do we really want to raise the price of water (or increase taxes elsewhere)? Do we want to do this to pay for extension of the central sewer service if people are going to pull up stakes and leave because their homes are flooding several times a

[28] For discussion see Flavelle, "Miami Will Be Underwater Soon: Its Drinking Water Could Go First." On affordability issues, see https://www.miamiherald.com/news/business/real-estate-news/article229029784.html.

year? And, probably the clincher, How likely are we to still be in these jobs after the next election if we agree to raise $3.6 billion in extra tax revenue? Perhaps I am being unduly cynical but, discretion being the better part of valor and all that, I suspect the outcome might be a decided punting of the ball into touch: a committee or two is set up to look into "feasibility" and so on and so forth until a few election cycles are successfully negotiated and it eventually becomes someone else's problem.

The same issues of cost are reiterated when we look at the drinking water question. It is not as if the problem of saltwater intrusion is technologically insurmountable. It is just that the solutions required are going to be rather costly. One possibility, for example, is desalination. In 2013, a desalination facility opened west of Hialeah.[29] The plant works on the principle of what is known as *reverse osmosis*. Brackish water is pulled up from 1000 feet beneath the surface and then pushed through a series of plastic membranes at a pressure of as much as 200 pounds per square inch. There are at least two drawbacks to this. First, it consumes a lot of energy: 5000 kilowatt-hours of electricity per million gallons of water. The worry is that widespread use of desalination will actually exacerbate climate change, which is, of course, responsible for the sea-level rise that created the need for the desalination plants in the first place. Second, it is a very expensive process. The plant itself costs $55 million to build, and the water it produces costs two-and-a-half times as much as water obtained from the Biscayne Aquifer by standard means. Desalination may eventually be inevitable for Miami-Dade. But it is going to significantly push up the water bills of everyone.

It might also be possible to establish new inland wells, although where is not entirely clear—and these will only be viable until the ocean sweeps in across the Everglades from the Gulf of Mexico. We might build new desalination plants and/or new and more efficient water treatment plants. We might build new over-ground reservoirs capturing the plentiful south Florida rainfall. We might do this, or we might do that. But whatever we do will cost money. A lot of it. And it is money that will, ultimately, provide the coup de grâce to Miami.

Miami-Dade faces certain financial pressures, some of which have to do with previous environmental malfeasance. In 2013, after it had been caught unlawfully dumping more than 28 million gallons of wastewater in Biscayne

---

[29] For discussion, see Flavelle, "Miami Will Be Underwater Soon: Its Drinking Water Could Go First."

Bay, Miami-Dade entered into an agreement with the EPA to upgrade its wastewater collection and treatment facilities. This is estimated to cost somewhere in the region of $1.6 billion. In a recent budget, the Department of Water and Sewer released a budget estimating a further $13.5 billion would be required for current and future infrastructure improvements.[30] And the county's climate-related expenditures, of course, go far beyond merely protecting drinking water and include costs of new pumps, overhaul of the county's largely gravity-based flood defenses, raising roads, and so on. These are expenses that few counties could afford. And in a state where battles over who gets funded for what are the norm rather than exception, any help from the state capital of Tallahassee is likely to be muted.

Federal money is a possibility, of course, but is also unlikely. The issues arising with respect to federal aid seem to mirror those that emerged at the local level. Lawmakers in Washington may, not unreasonably, decide that investing huge amounts of money in order to save Miami looks uncomfortably like throwing good money after bad. Issues of equity also arise. Living in Miami is, ultimately, a choice—in the sense that living somewhere else is always a possibility—and why should someone from, say, the Midwest pay more taxes in order to fund that choice? Indeed, the idea of moving somewhere else seems to be part of the climate plan of action (if I may speak loosely) of the current US administration. As Secretary of State, Mike Pompeo, puts it in an interview with the *Washington Times*, "The climate's been changing a long time. There's always changes that take place. . . . Societies reorganize, *we move to different places*, we develop technology and innovation."[31] Perhaps it is difficult to read the attitude of future administrations into the present one. Nevertheless, it is still true that people, in general, do not like paying more taxes to fund the lifestyle choices of others. Their being forced to do so would be very unpopular, a fact that will not be lost on their representatives in the Senate and House. Thus, I doubt significant amounts of federal money will come flooding in (no pun intended) to save Miami.

It seems overwhelmingly likely, therefore, that the cost of saving Miami—should this be possible—will ultimately fall squarely on the shoulders of those who choose to live here. Like me. That is, perhaps, as it should be. But it is unlikely to work. The demise of Miami will ultimately be shaped and decided by money, by the ratio of investment to return. Specifically, it

---

[30] Flavelle, "Miami Will Be Underwater Soon: Its Drinking Water Could Go First."
[31] "Pompeo: We Will Do the Things Necessary as the Climate Changes," *Washington Times*, June 7, 2019 (italics are mine).

will be decided by how much people are willing to pay before they decide enough is enough. Imagine: it is another day in "paradise." Your septic tank is repeatedly overflowing, and your house is subject to habitual flooding. The drinking water, you suspect, is not what it was. (Perhaps it is your imagination, but you can't help but notice a distinctly brackish tang.) The County is asking you for $50,000 to hook you up to the centralized sewer system. At the same time, your water bill is going through the roof, thanks to infrastructure improvements implemented by Miami-Dade Department of Water and Sewer and the new desalination plant they want to build out west. Your property tax is also increasing significantly, so that your municipality can address local flooding issues occurring during the increasingly intense heavy rains. Don't even mention the increase in your flood insurance over the last couple of years. And you're lucky: some can't even buy flood insurance, and so can't get mortgages. You have been offered a job in another state. In these circumstances, you would really have to *love* South Florida to want to remain. More likely is that you will move somewhere else, in state or out of state. And if you don't, others in your situation will. Atlanta is already becoming the go-to place for Miami expats, along with the always fashionable Carolinas. You'll have trouble selling your house, of course, because of the flooding, effluent, and suspect water. Perhaps it will sell. Perhaps it won't. You put up a *For Sale* sign and leave. You're not the only one, of course. These signs start springing up in many places with people who have simply had enough of the increasingly burdensome cost of paradise. But one less household is one less source of property tax. And one less person translates into less sales tax (and there is no state income tax in Florida). The State, and the County, therefore will have less money to spend on things like new inland wells, desalination plants, and extension of the centralized sewer system. The problems, accordingly, become worse. And more people leave. And tax revenues fall. And the problem will become still worse. You'll be replaced, of course, for the time being, by others from colder climes longing for a taste of the dolce vita. But eventually word will get out. The replacements will dwindle. And Miami will no longer have enough money to do what needs to be done to save itself.

You Can Hide from the Wind

Sea-level rise is not something that will happen in the future. It is here already. You can see it. This is what sea-level rise, in place such as Miami, looks like. It is not an angry ocean sweeping in triumphantly over the Columbus Day Regatta. It is wading through a foot or two of water during a king tide on Miami Beach: that's obvious. But it is also the clogged septic tank leaking foul effluent on to your carefully manicured lawn. It is a new well that is being sunk in western Miami-Dade, away from the ocean. It is an old well abandoned to saltwater intrusion in Broward. It is the new desalination plant being built by the County. It is the conflict between the County, who want to raise the roads of Miami Beach to stop them from flooding, and the residents who adjudge, correctly, that this will merely divert floodwater from the roads onto their properties.

This is what the death spiral of Miami will look like. Miami will become unlivable long before it sinks. This is the death of thousands of *For Sale* signs and empty houses, increasing all the time, as people decide that enough is enough and move to Atlanta. And few and fewer come in to replace them. This is the death of sky-high utility prices and forever dwindling tax revenues. Miami does not go out with a bang. It goes out with a whimper. Perhaps it is the sort of whimper you might utter when you stroll out into your beautiful garden one glorious sunlit morning and step up to your ankles in a pool of festering septic tank emissions. Actually, that might be more like a scream or yell than a whimper. I don't know. But we'll soon find out. Scream, yell, or whimper: it is a crappy way to fall from paradise.

In Florida, there is an old, hurricane-related saying: *you can hide from the wind but you have to run from the water*. If the problem of climate change were limited to sea-level rise, it wouldn't really be much of a problem at all. Not for those of us in the developed world, at any rate. We could simply run from the water, as Mike Pompeo seems to advise: solve our problem through the simple expedient of moving elsewhere. Easier for some people than others, of course, but certainly doable, especially for the principal targets of this book—denizens of the developed world. We care, perhaps disproportionately, about sea-level rise because, and to the extent that, it affects us, and does so in obvious ways. Its impact is easy to see and quantify: we can see it in the streets and quantify it by the value of the real estate it erases. After all, we humans have always cared most about ourselves, and caring for things other than ourselves is, for us, a bit of an ask. Sea-level rise, however, is just one facet of climate change, and far from the worst. We can hide from the wind and run from the water. But we can neither run nor hide from climate change. Climate change has many guises and, if unchecked, will get us in one way or another. It is to this that we now turn.

3

Welcome to Venus?

It's Hot in Here

The rising sea level that will eventually cause the demise of Miami is just one of the more obvious symptoms of something that can no longer be doubted: the planet is getting warmer. Ocean levels are rising not just because warmer water expands, and has more volume than colder water but, more importantly, because the Arctic, Antarctic, and Greenland ice caps are melting. And when these ice caps comprise land-based ice and glaciers, the meltwater flows into the ocean, raising its level. Despite some skepticism on Capitol Hill—"Where is that good old-fashioned global warming? I hope it comes back soon," once opined President Trump during an unseasonably cold spell in the Midwest—no one who knows anything about the facts on the ground doubts that the earth has recently been getting warmer.

Globally, 19 out of the last 20 years have been the warmest on record—since global temperature records began in 1880. According to NASA, 2019 was the second hottest year ever, with global temperatures (land and sea combined) at 0.95°C (1.71°F) above the twentieth-century average. The year 2016 was the warmest year on record, sporting global temperatures 1.10°C (1.98°F) above the twentieth-century average.[1] This beat the previous record set by 2015. The year 2017 saw temperatures of 0.91°C (1.64°F) above the twentieth-century average.[2] The 2018 global temperature was 0.79°C (1.42°F) above the twentieth-century average.[3] The slight drop in 2017–2018 is attributed to the lack of a pronounced *El Niño* phenomenon in those years. El Niño is a complex series of climatic changes that begins with the arrival of unusually warm, and nutrient-poor, Pacific water off Ecuador and Peru. It usually begins around December and will continue for several months thereafter. El

---

[1] National Oceanic and Atmospheric Administration, https://www.ncdc.noaa.gov/sotc/global/201604.

[2] National Oceanic and Atmospheric Administration, https://www.ncdc.noaa.gov/sotc/global/201901.

[3] NASA, https://climate.nasa.gov/news/2841/2018-fourth-warmest-year-in-continued-warming-trend-according-to-nasa-noaa/.

World on Fire. Mark Rowlands, Oxford University Press. © Oxford University Press 2021.
DOI: 10.1093/oso/9780197541890.003.0003

Niño generally tends to raise global temperatures by around 0.11°C (0.2°F). While 2017 and 2018 were slightly cooler than 2016, they were, in fact, the hottest years on record without an El Niño event. One of the goals of recent environmentalist efforts—reflected in *The Paris Agreement*—is to try to hold global temperature rises to "well below" 2°C above preindustrial levels. In some years, we are already more than halfway there.[4] Indeed, the above temperatures are global: they are averages of temperatures recorded both on land and over the oceans. Since temperatures on land are usually much higher than temperatures over the ocean, it is likely that the Paris Agreement's 2°C threshold has already been breached on land. Whether or not this is the case, not only is the warming trend clear, it is increasing in pace. Most of the roughly 1°C jump between late nineteenth century and today has occurred during the past 35 years. NASA has a nice computer mock-up here, accompanied by some suitably ominous music.[5]

The expression "global warming" has dropped out of fashion in recent years and has been replaced by the less specific "climate change." The alleged rationale for this is that the effects of climate change can be multifaceted and unpredictable. Not everywhere in the world will get warmer. In fact, some places might actually become colder. True. But, as far as I can see, this is a clear case of pandering to the most unreasonable skeptic imaginable. Does anyone really think that global warming entails that everywhere on the globe, without exception, must become warmer, and if it doesn't then global warming is not real? Of course, it entails no such thing. What it does mean is that the *average* global temperature—temperature averaged out all across the globe—is increasing. Nor does it have to increase every year. There can be local fluctuations—as we saw, 2017–2019 were not quite as warm as 2016, probably because of the lack of an El Niño effect. But, if global warming is real, then, *as a general trend, temperature averaged out across the globe must be increasing.* The data we have accrued since 1880 certainly suggests this is the case.

Moreover, abandoning the expression "global warming" has, arguably, been harmful to the extent it has allowed some skeptics to deny that the world is getting warmer in favor of the claim that some parts are getting warmer

---

[4] The Paris Agreement adds, "and to pursue efforts to limit the temperature increase even further to 1.5 degrees Celsius." https://unfccc.int/process-and-meetings/the-paris-agreement/the-paris-agreement. To see just how close we already are to the 1.5°C threshold, see https://climate.nasa.gov/vital-signs/global-temperature/.

[5] NASA, https://www.nasa.gov/press-release/nasa-noaa-data-show-2016-warmest-year-on-record-globally.

and others are not. Some of these skeptics are rather powerful. Thus, in an interview with Piers Morgan on *Good Morning Britain*, President Trump claimed: "I believe that there's a change in weather and I think it changes both ways." He added: "Don't forget, it used to be called global warming, that wasn't working, then it was called climate change, now it's actually called extreme weather because with extreme weather you can't miss."[6] The expression "global warming" was, in fact, working perfectly well. The phenomenon we are talking about is a global one: it pertains to average global temperatures. And it is a warming phenomenon. Some parts will get warmer, and some parts may indeed get colder. But, taken as a whole, the earth—the *globe*— is getting *warmer*.[7] I have no compunction in using the expression "global warming" and will continue to do so however much this will grind the gears of some. I shall also talk of "climate change," but the way I am using this expression means—apart from a few minor contextual differences here and there—the same thing as "global warning."

Skepticism about Global Warming, Part 1

We live in an age of group polarization abetted both by the mainstream and social media. In this age, there is an unfortunate tendency to portray those who disagree with one's views as idiots who should, perhaps, be "canceled." I find this regrettable and very uncivilized. I shall try to convince you of the reality of anthropogenic climate change or global warming, where "anthropogenic" means caused by humans or of human origin. In other words, I shall try to convince you that global warming is real and we are causing it. Those who are already convinced can skip to the next chapter. But I don't think climate change deniers are, in general, idiots. Don't get me wrong: some of them are. But a respectable case can be made out for climate change—global warming—*skepticism*. Roughly—as we shall see in a minute, this view can take several forms—global warming skepticism is the view that we don't have enough evidence to conclude, definitively, that global warming is occurring. By describing this case as "respectable," I don't mean to imply that it is true.

---

[6] As reported in the *Daily Mail*, June 4, 2019, https://www.dailymail.co.uk/news/article-7104153/ Piers-Morgan-gets-interview-Trump-hes-UK.html.

[7] Compare this with the issue of sea-level rise. As we saw in Chapter 2, if the Greenland Ice Sheet melts, sea levels around, say, Iceland, might drop, due to the loss of gravitational attraction hitherto exhibited by the sheet. Does this mean that sea-level rise is not real? I wish it were that easy.

It's not. But what I do want to insist is that there is a case for global warming skepticism that is not unreasonable or irrational. I am going to make this case in as sympathetic a way as I know how. And then I shall argue that it does not add up. Skepticism about climate change, or global warming, is not, necessarily, stupid. But it is, nevertheless, false.

Skepticism about global warming can take several forms. One form denies that the earth is really getting warmer (or claims that it is not doing so abnormally). Another form accepts that it is getting warmer, but it denies that this warming is *anthropogenic*. That is, it denies that the warming is being caused by the activities of humans. There are, as we shall see, further forms, and we shall look at them all in due course. But let's begin with the first sort, which denies that the earth is really getting warmer. This first sort of skepticism will appeal to either to the *paucity* or *lack of accuracy* of available global temperature records. Temperature records come in two forms: *surface* and *satellite*. Surface measurements derive from a large collection of thermometers that are dotted around the globe, both on land and on the surface of the ocean (on ships or, more commonly, on buoys). Surface records date back to 1880. Satellite measurements are collected from a number of satellites that circle the earth above the poles. These satellites examine the *troposphere*, the region of the earth's atmosphere extending up to 10 kilometers above the ground. They use microwave instruments that measure how much heat is emitted by tropospheric oxygen molecules, which can in turn be used to calculate air temperature. Satellite records date back to 1979.

The first type of skeptic about global warming can point out that even 140 years, the age of surface records, is but the blink of an eye in the life of a planet. This is, of course, true. Are there no other records, one might ask? Ones that we might be able to use to reconstruct global temperatures prior to 1880? There are, in fact, such records, and these provide us with a longer-term picture of the global temperature profile of the plant. However, the picture they paint does not provide a convincing response to the skeptic. These other records consist in what are known as *proxy* measurements and include measurements of tree rings, coral growths, and the variation of isotopes in ice cores. The evidence provided by these phenomena is necessarily sparse, geographically speaking, but the measurements do converge on a definite picture. It is a picture of a fairly regular series of temperature peaks and troughs. If we look back for a period of 2000 years, then the temperature of the planet today corresponds to a distinctive peak, larger than anything we have seen in this time frame. However, if we look back further—800,000 years—then what we see is a series of peaks and troughs, roughly about 100,000 years apart, corresponding to a sequence of glacial and interglacial periods.

Due to fluctuations in its orbit, in particular wobbles in its orientation to the sun, the earth regularly cycles between glacial and interglacial periods. These wobbles are known as Milankovic cycles, after the Serbian astronomer Milutin Milankovic, who first hypothesized their role in the glaciation cycle. Interglacial periods correspond to periods of more intense summer solar radiation in the Northern Hemisphere. At present we are in an interglacial phase and, accordingly, our current global temperature is represented as one of the peaks on a graph of temperature change on a geologic time scale. NASA has a useful set of these graphs here.[8] At first glance, there does not seem to be anything particularly distinctive about our peak. Indeed, the previous peak of c. 100,000 years ago was somewhat higher. In short, the global warming skeptic's case can be put like this: surface and satellite readings are too recent to conclusively establish that we are entering into a period of distinctive, or unprecedented, global warming. And proxy measurements do not establish that we are in anything more than just another interglacial. This is not an unreasonable case, and it cannot simply be dismissed.

The skeptic can further strengthen his case by pointing to discrepancies between the surface and satellite records. Since 1979, surface records report a globally averaged warming of 0.16°C per decade. Satellite records, however, report an average warming of 0.11°C per decade. Thus, the skeptic can argue, in addition to the short-lived character of surface and satellite records, there is also a question about their mutual consistency. Why, the skeptic might, not unreasonably, ask should we place such store in evidence that is so transitory and not entirely consistent?

Response to the Climate Change Skeptic, Part 1

The first response to the skeptic involves establishing that something unusual is, in fact, going on today: something entirely distinctive about our current period of warming. This the sheer *rate* of change of global temperature. As the proxy records show, the earth regularly cycles between glacial and interglacial periods. It is also true that glacial periods can end abruptly,[9]

---

[8] NASA, https://earthobservatory.nasa.gov/features/GlobalWarming/page3.php.
[9] See, for example, the NOAA (National Oceanic and Atmospheric Administration) website, https://www.ncdc.noaa.gov/abrupt-climate-change/Glacial-Interglacial%20Cycles.

and we shall look at some reasons for this later. Nevertheless, there is something about today's warming trajectory that is a little "off." On a geological timescale, what counts as "abrupt" encompasses thousands of years. For example, the *last glacial maximum*—the most recent time of peak ice coverage of the planet—occurred between roughly 26,000 and 20,000 year ago.[10] The "abrupt" ending of the last glacial period took place, largely, between 20,000 and 14,000 years ago. That's an abrupt 6000 years. However, much of the change in temperature from 1880 to today has occurred in the last 35 years. That is an entirely different sense of abrupt. In environmental systems, rate of change is far more important than change itself. If these sorts of change occurred over thousands of years, then the earth and its biota would have time to adapt. But when the change occurs in decades, there is no possibility of adaptation, and environmental catastrophe potentially beckons. The current rate of global temperature change is unprecedented. Nothing in the proxy global temperature record looks anything like this. The skeptic appeals to the proxy record to argue that what we face today is, at most, *just another interglacial*. The first response to the skeptic is to point out that the change in global temperature we are seeing today is nothing like any interglacial we have ever seen before.

The second response to the skeptic is to point out that surface and satellite temperature records are not the only evidence we have that the earth is entering a period of unprecedented climatic change—of the warming variety. There are also theoretical reasons, and given these reasons, the rapid change in global temperature reflected in surface and satellite readings is just what we should expect.[11] The theoretical reasons pertain to the amount of carbon dioxide (CO_2) and other *greenhouse gases* (GHGs) in the atmosphere. At the time of writing, the earth's atmospheric carbon dioxide is 416.18 parts per million (ppm).[12] In 1960, this figure stood at 320 ppm. In preindustrial times, atmospheric carbon dioxide ranged from 200 to 280 ppm—around 200 during glacial and going as high as 280 during interglacial periods. The main reason for believing the planet is in an unprecedented phase of global warming is that we know that certain gases—such as carbon dioxide, methane, nitrous oxide, and *chlorofluorocarbons* (CFCs)—have a *greenhouse*

---

[10] P. Clark, A. Dyke, J. Shakun, A. Carlson, J. Clark, B. Wohlfarth, J. Mitrovica, S. Hostetler, and A. McCabe, "The Last Glacial Maximum," *Science* 325 (2009): 709–714.

[11] Stephen Gardiner has also made this point. See his "Ethics and Global Climate Change," *Ethics* 114 (2004): 555–600, 567–568.

[12] https://www.co2.earth/. Accessed May 30, 2020.

effect: they trap heat near the earth's surface. We not only know *that* they do this. We also know *why* they do this.

With respect to greenhouse potential—the potential of a gas to trap heat near the earth's surface—two's company, three's a calamity. At least a potential one.[13] The bulk of our atmosphere comprises nitrogen (N_2) and oxygen (O_2). Nitrogen and oxygen molecules are made up two atoms. Molecules with two atoms can vibrate only in one way: by moving closer together and then further apart. Carbon dioxide, however, is made up of three atoms, one carbon atom and two oxygen atoms. A three-atom molecule vibrates in more complex ways—called *vibration modes*. The greater the number of vibration modes of a molecule, the more likely that molecule is to interact with passing waves of electromagnetic radiation. This is why carbon dioxide absorbs infrared radiation while nitrogen and oxygen molecules do not. Carbon dioxide also emits infrared radiation, but this is likely to be picked up by neighboring carbon dioxide molecules which are vibrating in the same kind of way. The result is that, whether absorbed or emitted, infrared radiation—heat, as most of us know it—is kept near the surface of the earth. This is the greenhouse effect. Methane is made up of five-atom molecules—a carbon atom attached to four hydrogen atoms—and these can enter into even more complex vibration modes. And this makes it even better at trapping heat.

The strongest reason for thinking that we are entering an era of unprecedented global warming lies in the role we know carbon dioxide (and methane, and other GHGs) to play in trapping heat near the surface of the earth. Graphs of the increase of atmospheric carbon dioxide almost perfectly match graphs of global temperature increase. We know that this should be so and why it should be so. Global temperature records are compelling not because they are comprehensive or complete but because they reflect this underlying theoretical connection between atmospheric carbon dioxide (and other GHGs) and global temperature.

The other strand of the skeptic's argument appealed to the discrepancy between surface and satellite records. The former measurements indicate a global temperature increase of 0.16°C per decade, while the latter indicate an increase of (only) 0.11°C per decade. We should not, however, attach too much significance to this discrepancy. The first thing to remember is that while there is a 0.5°C difference between the estimates, both agree that global temperatures are increasing decade on decade. Secondly, the reasons for

---

[13] More than three is, I believe, known as a "clusterfuck."

the discrepancy are well understood. Satellites have trouble picking up the heat of the El Niño phenomenon along the equator, where the warmer water condenses too high for satellites to pick up. As a result, El Niño is registered only as warmer areas north and south of the equator.[14] Third, more recent and more sophisticated satellite datasets have reduced the discrepancy between satellite and surface records to almost nothing.[15] Indeed, these new satellite datasets show slightly more warming since 1979 than the surface records. It is difficult, therefore, to read too much into any (alleged) divergence between surface and satellite records. Both concur that the earth is warming, and both are quickly converging on the question of by how much.

As we have seen, another form of skepticism about global warming accepts that the earth has recently been getting warmer but denies this warming is *anthropogenic*. That is, it denies that human activity is the source or cause of this warming, or denies that we have enough evidence to say unequivocally that the warming is caused by humans. However, in effect, we have already answered this form of skepticism. This form of skepticism, in effect, misunderstands the nature of the evidence for global warming. The best evidence for global warming is the connection between three-or-more-atom gas molecules (such as carbon dioxide and methane) and the trapping of heat due to the vibration modes of these molecules. These theoretical reasons can be supplemented or supported by the historical global temperature records. But the main work in the case of global warming is being done by these theoretical reasons. We know exactly why more and more carbon dioxide is finding its way into the atmosphere: the burning of fossil fuels by humans. We know exactly why more and more methane is finding its way into the atmosphere: the vast number of cattle raised by humans for food, among other things. The best evidence for global warming is, at the same time, evidence for this warming being anthropogenic. In short, we know global surface temperatures have been rapidly increasing. We know, also, that atmospheric carbon dioxide and methane have been increasing at the same kind of rate during the same period. We know that this latter increase is not an accident: we put it there. And we know that an increase in atmospheric carbon dioxide and methane should lead to an increase in global surface

---

[14] https://www.carbonbrief.org/explainer-how-surface-and-satellite-temperature-records-compare.

[15] Carl Mears and Frank Wentz, "A Satellite-Derived Lower-Tropospheric Atmospheric Temperature Dataset Using an Optimized Adjustment for Diurnal Effects," *Journal of Climate* 30, no. 19 (2017): 7695–7718, https://journals.ametsoc.org/doi/full/10.1175/JCLI-D-16-0768.1

temperatures. If you put those four things that we know together, the case for specifically anthropogenic global warming seems inescapable.

A Little More Skepticism: The Argument from Paleoclimate

There is still a card or two that the climate change skeptic might wish to play. The bad news for the skeptic is that I think we're getting near the bottom of the deck, and this is probably the last available hand. The good news is that it is, perhaps, the best—and most subtle—argument yet for climate skepticism. This type of skepticism I have in mind acknowledges that carbon dioxide and other GHGs are drivers of, or contributors to, global warming. But it denies that these are the *only* drivers of global warming, and it denies that these are the *most significant* drivers of global warming.

The proposed justification for this version of climate skepticism lies in periods of the earth's history when atmospheric carbon dioxide was much higher than it is now and yet temperatures were, allegedly, not that different from the current one. For example, during the Jurassic period (201–145 million years ago), the atmosphere is estimated to have contained roughly 1950 ppm of carbon dioxide, compared to a little over 400 ppm now. The Cambrian period (541–485 million years ago) saw levels of 7000 ppm—mainly due to volcanic activity.[16] These figures are sometimes used by climate change skeptics, who argue it shows that atmospheric carbon dioxide is not a significant driver of global warming. If carbon dioxide levels can be so much higher, while global temperatures remain not all that much different from today, this shows that there are other factors driving global temperature, more important than levels of atmospheric carbon dioxide.

This argument is an interesting one, but I do not think it is ultimately going to work. The first thing to note is that estimating temperatures in prehistoric times is quite difficult. All we have to go on are the concentrations of certain isotopes in rocks and sediment. The resulting estimates are likely to be very imprecise. Nevertheless, what estimates we do have for global

---

[16] Daniel Rothman, "Atmospheric Carbon Dioxide Levels for the Last 500 Million Years," *Proceedings of the National Academy of Sciences* 99 (2002): 4167–4171.See especially D. van Der Meer, R. Zeebe, D. van Hinsbergen, A. Sluijs, W. Spakman, and T. Torsvi, "Plate Tectonic Controls on Atmospheric CO_2 Levels Since the Triassic," *Proceedings of the National Academy of Sciences* 111 (2014): 4380–4385.

temperatures during the Jurassic period are notably higher than present global temperatures: probably around 5°C, but anywhere from 2°C to 8°C higher.[17] This is, actually not far off from what we would expect given the estimated levels of carbon dioxide in the atmosphere at that time.

The Cambrian is a different case and needs to be assessed differently. The average global temperature during the Cambrian is not known with any precision, but the consensus is that the planet was somewhat warmer then than it is today. There is no evidence of ice at the poles during this period. However, the most important thing to bear in mind when assessing paleoclimate in the Cambrian is that the sun was significantly cooler than it is today. (The same is true, to a lesser extent, for the Jurassic). When the earth first formed, the sun produced only around 70% of the level of total solar irradiance that it produces today.[18] A gradual strengthening of irradiance in the first half of its life is part of the natural history of most stars. The luminosity of our sun has been slowly increasing in the intervening periods. Without a significant greenhouse effect, the earth would have been a frozen ball in those early days of its history. Now, however, with our sun significantly warmer, current levels of GHGs are proving problematic. If you combine both factors—the lesser solar irradiance of the sun in the Cambrian and the higher global temperatures during the Jurassic—the most reasonable conclusion to draw is that these historical temperatures assessments do not contradict the idea that atmospheric carbon dioxide is an important driver of global warming. If anything, they seem to support this idea.

At this point, the skeptic *might* be willing to concede the Cambrian and Jurassic but is likely to dig his or her heels in at the Late Ordovician (460–443 million years ago). At that time, atmospheric carbon dioxide levels that were thought to be very high, roughly 5600 ppm, coincided with a glaciation event so severe that it resulted in one of the greatest marine extinctions ever. How, the skeptic might legitimately ask, could such high atmospheric carbon dioxide levels be compatible with this sort of severe planetary glaciation if carbon dioxide is such an important driver of global warming?

There are several problems with this appeal to the Late Ordovician. First, coming not long after the Cambrian—on a geological time scale at least— the earth at that time was bathed in the light of a nonnegligibly colder sun

---

[17] "Paleoclimate," *Encyclopedia Britannica*, https://www.britannica.com/science/Jurassic-Period/Paleoclimate.

[18] G. Feulner, "The Faint Young Sun Problem," *Reviews of Geophysics* 50, no. 2 (2012): RG2006, 1–29.

(roughly 4% cooler during the late Ordovician than it is today).[19] Second, more recent data have cast serious doubt on the idea that carbon dioxide levels were really that high during the time of the glaciation. A 2009 study examined the level of strontium isotopes in sediment form the Late Ordovician. Strontium is produced by the weathering of rock, and weathering is a process that removes carbon dioxide from the atmosphere. The level of strontium in the sediment, therefore, is a good indicator of how much carbon dioxide is being removed from the atmosphere through weathering. On the basis of this, the study estimates a level of carbon dioxide in the atmosphere of below 3000 ppm, which, in combination with a cooler sun, might well have been enough to initiate glaciation.[20] Third, the late Ordovician saw a widespread explosion of terrestrial plant life. These early mossy plants greatly accelerated rock weathering, as well as drawing down carbon dioxide from the atmosphere themselves. They also supplied nutrients such as phosphorus to the oceans, which resulted in a further explosion of plankton activity, which further drew down carbon dioxide from the atmosphere and acted as carbon sinks as their remains fell to the seabed. Consequently, more recent estimates place the level of atmospheric carbon dioxide at between 1000 and 2300 ppm[21]—a level that, in combination with a 4% weaker sun, would certainly trigger a glaciation event. The paleoclimate record of the late Ordovician, therefore, probably does not contradict the close relation we know to hold between atmospheric carbon dioxide and the greenhouse effect.

This last type of skeptical argument—which seeks to deduce climate skepticism from the paleoclimate record—is, therefore, unlikely to work. Moreover, its central contention—that while carbon dioxide is a driver of climate change it is not the only or most important driver—is, when you think about it, a truism. Is there really anyone who would want to claim that atmospheric carbon dioxide and the other GHGs are really the most important determinants of the climate? As a general claim, this is obviously false.

---

[19] Possart et al. have argued that given a 4.5 diminution in solar luminosity, permanent snow cover of 60% of the planet is compatible with atmospheric carbon dioxide levels being 10× current levels. P. Poussart, A. Weaver, and C. Barnes, "Late Ordovician Glaciation under High Atmospheric CO_2: A Coupled Model Analysis," *Paleoceangraphy* 14, no. 4 (1999): 542–558.

[20] Seth Young et al., "A Major Drop in Seawater 87Sr/86Sr during the Middle Ordovician (Darriwilian): Links to Volcanism and Climate?" *Geology* 37, no. 10 (2009): 951–954, https://pubs.geoscienceworld.org/gsa/geology/article-abstract/37/10/951/103964/a-major-drop-in-seawater-87sr-86sr-during-the?redirectedFrom=fulltext.

[21] Richard Pancost et al., "Reconstructing Late Ordovician Carbon Cycle Variations," *Geochimica et Cosmochimica Acta* 105, no. 15 (2013): 433–454, https://www.sciencedirect.com/science/article/pii/S0016703712006862.

The sun, in general, is always the most important driver of global warming. One day, for example, our sun will rapidly expand into a red giant, absorbing the orbits or Mercury, and perhaps even Venus, within it. This, we can safely conclude, will have somewhat calamitous global warming effects: the earth will be reduced to a molten ball of metal, and the level of atmospheric carbon dioxide when this happens will be completely irrelevant. However, that day, thankfully, is not today. The sun is not going to change very much for the next few million years or so, certainly not in the kind of time frame we should be concerned about—a time frame measured in decades or centuries rather than millions of years. Given that the total solar irradiance will remain more or less fixed for this foreseeable future, and so play no significant role in driving up or driving down global temperatures during this time, the most important de facto drivers of global temperature change today, and for the sort of time span we need to worry about, are the GHGs we are putting into our atmosphere.

How Bad Will It Get?

We know the earth is warming, and we know this warming is anthropogenic. We also know, in general terms at least, how to stop it. But it is not clear that we have the stomach to do so. Not yet, anyway. In such circumstances, an obvious question arises: How far will this go? Again, this depends on how we behave over the rest of this century. Projections of future warming potential require the assessment of various, largely uncertain variables. These include uncertainties over the sensitivity of the climate to increased volumes of atmospheric GHGs, over the quantities of gases emitted and ratios of different types of GHG emissions, over the success of future technologies such as carbon capture and sequestration,[22] future population growth patterns, and a myriad of other factors. To try to understand the range and variety of possible future emissions, energy system modelers use what are known as *integrated assessment models*, which simulate future energy emissions and technologies. These models supply emissions scenarios that are then fed into climate models, which simulate how the climate might change in the future under a variety of emissions scenarios.

---

[22] See Chapter 5 for discussion.

For its Fifth Assessment Report (AR5) published in 2013, the Intergovernmental Panel on Climate Change (IPCC) employed four different model scenarios, known as *representative concentration pathways* (RCPs).

- RCP 2.6 is a high-end mitigation model. It assumes carbon dioxide emissions will be drastically reduced by 2040, and reduced to zero by 2080. In roughly the same period, methane emissions will be significantly reduced, although not to zero, and nitrous oxide emissions will, basically, flatline. Result: Global temperature increase < 2°C (*Medium confidence*).
- RCP 4.5 is a lower-end mitigation model. It assumes that carbon dioxide and methane emissions will both be roughly halved in the medium term (i.e., by 2080), and that nitrous oxide emissions will flatline in roughly the same time frame. Result: Global temperature increase > 2°C by 2100 (*Medium confidence*).
- RCP 6 is a low-end business-as-usual model. It assumes carbon dioxide emissions will peak around 2060, and then decline, although not to today's levels. Methane and nitrous oxide will undergo modest increases during the same period, although methane emissions will start to decline post 2070. Result: Global temperature increase >2°C by 2100 (*High confidence*).
- RCP 8.5 is a high-end business-as-usual model. It assumes all three types of GHG emissions will continue to increase, the result of high population and low rate of technological progress. Result: Global temperature rise between 3.5°C and 4.9°C by 2100 (*High confidence*).[23]

Initially (c. 2010–2011) these scenarios were designed with a view to none being any more, or less, plausible than the others. However, generally speaking, RCP 8.5 is regarded, today, as somewhat less likely than the others due, in large part, to what many people think are its unrealistically high assumptions about the level of coal use during the rest of this century. Nevertheless, the demise of RCP 8.5 is by no means a done deal. It is just as compatible with current trajectories as the alternatives. Moreover, a recent

---

[23] The numbers derive from what is known as the *radiative forcing* assumed in each scenario. Radiative forcing is a function of the discrepancy between the amount of solar radiation trapped near the surface of the planet versus the amount escaping into space. The assumptions deployed in each scenario equate to a certain radiative forcing value. For these figures, and the respective confidence they inspire, see International Panel on Climate Change, Fifth Assessment Report (AR5), Chapter 2, "Future Climate Changes, Risks and Impacts," https://ar5-syr.ipcc.ch/topic_futurechanges.php.

published study by the Nobel Prize–winning economist William Nordhaus and colleagues argued that there is currently a 35% chance of RCP 8.5 becoming actualized—that is, of the world exceeding the RCP 8.5 trajectory by the end of the century.[24]

In 2017, RCPs were replaced by shared socioeconomic pathways (SSPs). These are more complex models, integrating additional population sets, and incorporating different assumptions about economic growth and other socioeconomic factors. RCP 2.6, 4.5, 6, and 8.5 were replaced by SSPs 1.9, 2.6, 3.4, 4.5, 6, and 8.5. But the overall picture remained very much the same. As with its RCP counterpart, SSP 8.5 represents a high-end business-as-usual scenario. The SSP 1.9 represents an extremely high-end mitigation scenario. And there are various stages of in-between. The graphs plotting SSP trajectories closely mirror their RCP forebears.[25]

Let's take a concrete number, in the middle of the range licensed by RCP 8.5 and also, on some interpretations, within the upper range of RCP 6: 4°C or just over 7°F. A middle-range business-as-usual scenario will entail something like this kind of global temperature increase by 2100. More precisely, prior estimates had assigned this increase as 62% probability under a middle-range business-as-usual scenario. However, more recent estimates now assign this increase a 93% probability.[26] Of course, a 93% probability of an increase of 4°C also entails somewhat lower probability of an increase of 5°C, and so on. But let us work with 4°C since it is largely uncontroversial.

It is generally accepted that 4°C would be bad. Very bad, according to a report by the World Bank.[27] The World Bank is not an institution prone to hyperbole, and the language it uses in the report is measured and cautious. Nevertheless, it seems the consequences we should expect include extreme heatwaves: the sorts of heatwaves we used to see only every hundred years or so would now, in many places, occur during most summers. A global increase of 4°C would not be evenly distributed. It will be much higher on land. And land is the place we live. Increases of 6°C–10°C should be expected in in the Mediterranean, North Africa, Middle East, and parts of the United

[24] P. Christensen, K. Gillingham, and W. Nordhaus, "Uncertainty in Forecasts of Long-Term Economic Growth," *Proceedings of the National Academy of Sciences* 115, no. 21 (2018): 5409–5414.

[25] B. O'Neill, E. Kriegler, K. Riahi, K. Ebi, S. Hallegatte, T. Carter, R. Mathur, and D. van Vuuren, "A New Scenario Framework for Climate Change Research: The Concept of Shared Socioeconomic Pathways," *Climate Change* 122, no. 3 (2014): 387–400.

[26] P. Brown and K. Caldeira, "Greater Future Global Warming Inferred from Earth's Recent Energy Budget," *Nature* 552, no. 7684, 2017: 45–50.

[27] The World Bank, "Climate Change," https://www.worldbank.org/en/topic/climatechange.

States. There are likely to be malign effects on agriculture, water resources, and human health, the result of drought, famine, and spread of newly emerging pathogens. Massive displacement of populations will be likely. Severe disruption of global trade is also inevitable. Sea-level rises of a meter or more will likely devastate vulnerable coastal cities in Bangladesh, India, Indonesia, Madagascar, Mexico, Mozambique, the Philippines, Venezuela, and Vietnam. Not to mention Miami. Many small islands will not be able to sustain their populations.

Perhaps even more worrying is what reports such as this might, possibly, miss. The key to understanding how bad the planet will eventually get depends on whether you think the relation between anthropogenic GHGs and global temperature is *linear* or *nonlinear*. The relation is linear if a given rise in an atmospheric GHG—say carbon dioxide—is directly proportional to a given increase in global temperature. Since I am, here, merely explaining the idea of a linear relation, the specific figures are not important. But suppose a rise of atmospheric carbon dioxide of 100 ppm is correlated with an increase of global temperature of 1°C. Suppose, further, that a rise of atmospheric carbon dioxide of 200 ppm is correlated with a 2°C global temperature rise. A 300 ppm rise correlates with a 3°C rise, and so on. If so, then the relation between atmospheric carbon dioxide and global temperature would be a linear one. If you look at graphs detailing the rise of atmospheric carbon dioxide and global temperatures, you will see a linear relation. In many ways this is reassuring. If the relation is linear, and always remains linear, we can predict, more or less, how much of a rise in global temperature will accompany a given rise in atmospheric carbon dioxide. Moreover, the identified correlation between atmospheric carbon dioxide and global temperature, coupled with a mechanism which explains how increased atmospheric carbon dioxide leads to increased global temperatures, provides the best evidence for anthropogenic global warming. Therefore, we should take seriously the idea that the relation between atmospheric carbon dioxide and global temperature is a linear one—and we should certainly keep our fingers crossed for this.

Once we take other GHGs into account, not just carbon dioxide but also methane, nitrous oxide, and CFCs—I suspect it is very likely that we are dealing with some sort of linear relationship between the concentration of these gases in the atmosphere and global temperatures. However, here is the rub: atmospheric carbon dioxide (and also methane and other GHGs) comes in two forms: *anthropogenic* and *nonanthropogenic*. Anthropogenic carbon dioxide is carbon dioxide that we put into the atmosphere, primarily through the burning of fossil fuels. Nonanthropogenic carbon dioxide is carbon dioxide that arrived in the atmosphere through other, nonhuman means. This adds a potentially worrying wrinkle. Even if we assume that the relation between atmospheric carbon dioxide *in general* and global temperature is linear, it does not follow that the relation between specifically *anthropogenic* carbon dioxide and global temperature is linear. Indeed, there are some considerations that suggest it is not or, at least, need not be.

Tipping Cascades and Runaway Warming

In a paper published in 2016, in the *Proceedings of the National Academy of Sciences*, Will Steffen and colleagues proffered what is, in effect, a nonlinear analysis of the relation between specifically *anthropogenic* atmospheric carbon dioxide and global temperature.[28] Anthropogenic carbon dioxide is carbon dioxide that we humans have put into the atmosphere. The worry developed by Steffen and colleagues is that this specifically anthropogenic carbon may act as a springboard, inducing further increases in global temperature, which then cause the further release of nonanthropogenic carbon dioxide, and other GHGs, which increase global temperatures still further. The end result will be a potentially disastrous positive feedback loop, continually driving up global temperatures even though the amount of specifically anthropogenic carbon dioxide in the atmosphere remains the same. This brings us to the nightmare scenario: the possibility of *runaway global warming*.

Suppose we take seriously the goal of the Paris Agreement to limit global temperature increases to no more than 2°C above preindustrial levels. Success may be unlikely, but we decide to give it our best shot. Our efforts assume a linear relation between atmospheric GHGs and global temperature. If we can limit atmospheric GHGs to such and such an amount, we believe, then we can limit temperatures increases to this threshold. We, in fact, do limit the amount of GHGs in the atmosphere to what we think is the required amount. Excellent, we might think: job done!

---

[28] W. Steffen, J. Rockström, K. Richardson, T. Lenton, C. Folke, D. Liverman, C. Summerhayes, A. Barnosky, S. Cornell, M. Crucifix, J. Donges, I. Fetzer, S. Lade, M. Scheffer, R. Winkelmann, and H. Schellenhuber, "Trajectories of the Earth System in the Anthropocene," *Proceedings of the National Academy of Sciences* 115 (2016): 8252–8259.

If Steffen and colleagues are right, this might not be enough to stop run-away global warming. Key to their analysis is the concept of a *tipping cascade*. An environmental tipping point is a point of drastic, irreversible change in the environment. A tipping cascade is a sequence of tipping points, where one tipping point has a series of effects that produces the next, which in turn has a series of effects that produces the next and so on. The imagined rise of 2°C would not be without certain effects. Specifically, we should expect sig-nificant melting of the Greenland Ice Sheet, of the West Antarctic Ice Sheet, of alpine Glaciers, and the almost total disappearance of Arctic summer ice. That's not good, you might think, but at least the damage has been done, and now we can get on with our lives in a world that has significantly less ice than it used to have but is still perfectly livable, for us if not for polar bears.

However, this conclusion would be premature. Ice is white, and white things reflect solar radiation back into space. This is known as an *albedo ef-fect*. The albedo of a substance is a measure of what proportion of solar radi-ation it reflects. The higher its albedo, the more radiation it reflects. Ice has a high albedo. As the ice sheets melt, they turn into oceans or earth, both of which, being darker than ice, have a much lower albedo, and so absorb more solar radiation, thus elevating global temperatures beyond the 2°C to which we thought we had limited ourselves. With its resulting lowering of albedo, 2°C was just a springboard for a further temperature increase. How much? We don't know. But it doesn't stop there.

The earth's oceans comprise great conveyor belts of thermohaline circula-tion. The Gulf Stream is, perhaps, the most famous of these, transporting warm water from the equatorial Atlantic to Northern Europe, keeping the western arm of this region surprisingly balmy for its latitude. However, the Gulf Stream is merely a stretch of the Atlantic Meriodonal Current. And the latter is merely a stretch in a planet-wide system of ocean currents. The water that reaches the Gulf Stream has made its way all the way across the Indian Ocean, past the Cape of Good Hope, and up through the southern Atlantic Ocean. This great Ocean conveyor belt is driven by two things: the *temperature* and the *salinity* of the water. The warm Gulf waters rise to the surface. When they reach a certain point of their northerly journey, they cool sufficiently to sink. The sinking water draws in warmer water over the top of it, and this is what sets up the conveyor of ocean currents. This pattern of rising and sinking is driven, in part, by temperature. But the salinity of the ocean water also plays an important role. When the water molecules of the ocean become heated, they expand. Extra space is created by this expansion into which salt molecules can fit. Since warmer water thus can

hold more salt and other molecules than cold water, it generally has a higher salinity. The higher the salinity of the water, the denser, and therefore the heavier
is that water. This greater density facilitates the sinking of the water when it
reaches colder climes. And the sinking of the water is essential to the setting up
of the thermohaline conveyor belt. When large quantities of ice melts and runs
off into the ocean, this additional freshwater reduces the salinity of the ocean
water. This impedes the maintenance of the world's ocean conveyor currents.
Evidence suggests that that the vigor of the Atlantic Meriodonal Current has
decreased by 15%–20% in the last 150 years.[29] And this, as well as having immediate effects on ocean ecosystems, may impact broader, land-based, ecosystems.
IPCC modelers have high confidence that the kind of scenario envisaged in
RCP 8.5 will result in a shutting down of the Atlantic Meriodonal Current.[30]

One can easily imagine the environmental impact on the land-based
ecosystems of western Europe occasioned by the slowing or shutting down
of the Atlantic Meriodonal Current. The region begins to look something
like Siberia. Vegetation will die. Dying vegetation gives off carbon dioxide,
increasing atmospheric levels of this gas, and so driving up temperatures
further. But the effects of the slowing of the Atlantic Meriodonal Current are
likely to be far more widespread than this. The great ocean currents function to transfer heat around the globe, preventing it from building up too
much in any given place. The Atlantic Meriodional Current takes heat out
of the tropics, transferring it to more northerly climes. With a slowing of the
current, not only will western Europe get colder, but the western equatorial
Atlantic region is likely to get hotter. Hotter and wetter, of course, generally translates into storms that are bigger, more frequent, and more violent,
which is, of course, bad news for me and 6 million other residents of South
Florida. But, more importantly, it will likely be bad news for ecosystems,
such as the Amazon rainforest, that are unable to keep up with this rapid
environmental change. The rainforest, already under pressure from the livestock industry, will likely suffer further damage, releasing more carbon dioxide into the atmosphere, and reducing the planet's ability to reabsorb—or
sequester—this carbon. This will drive up global temperature further. More

[29] D. Thornalley, D. Oppo. P. Ortega, J. Robson, C. Brierley, R. Davis, I. Hall, P. Moffa-Sancehz,
N. Rose, P. Spooner, I. Yashayaev, and L. Keigwen, "Anomalously Weak Labrador Sea Convection
and Atlantic Overturning during the Past 150 Years," *Nature* 556 (2018): 227–230. L. Caesar, F.
Rahmstorf, A. Robinson, G. Feulner, and V. Saba, "Observed Fingerprint of a Weakening Atlantic
Ocean Overturning Circulation," *Nature* 556 (2018): 191–196.
[30] International Panel on Climate Change, Fifth Assessment Report (AR5), Chapter 2: "Future
Climate Changes, Risks and Impacts."

generally, the great ocean currents are all connected: a slowing of one will inevitably have knock-on effects on the others. Patterns of weather will change as a result, and if these changes are significant and rapid enough, they will bring about the death of flora, and the fauna that depend on them, releasing yet more carbon dioxide into the atmosphere, and so driving up temperatures further.

At this point, the global temperature will continue to arc upward, and the world may be warm enough for further changes to occur. There will be further melting of sea ice, this time of Arctic winter sea ice and partial melting of the vast East Antarctic Ice Sheet. Again, less reflectance equals more absorption of solar energy, augmenting global temperature further. Of particular concern is the melting of the Siberian and North American permafrost. This will release large amounts of both carbon dioxide and methane. Methane is a very potent greenhouse gas, 28 times more efficient than carbon dioxide at trapping heat when measured over a 100-year time frame, and an alarming 72 times more efficient at trapping heat when measured over a 20-year time frame. (The difference is a result of the fact that methane breaks down much more quickly than carbon dioxide.) So the release of huge quantities of methane contained in permafrost is likely to significantly increase global temperature.

A tipping cascade is, essentially, a *positive feedback* loop. A given global temperature increase—like Steffen, we were working with 2°C—brings about certain changes which drive up the temperature further, which brings about further changes, which drive up the temperature further, which brings about further changes, which . . . and so on and so forth. The nightmare scenario is that this positive feedback loop becomes a runaway phenomenon. If this happens, the earth may be on its way to becoming Venus. But even if this doesn't happen, the earth may become much warmer than we anticipated, with all the attendant problems that this engenders. Steffen and colleagues estimate an eventual sea-level rise of up to 200 feet, for example. We thought we had done enough to keep global temperature increase to 2°C. But it turned out that 2°C had a life of its own, forcing itself ever higher and higher. 2°C was the genie in the bottle, and once it got out, it just became bigger and bigger and we could never put it back in again.

How likely is any of this to happen? The short answer is that, at the moment, we just don't know. Nevertheless, the account provided by Steffen and colleagues is persuasive, in part, because it incorporates factors that are already used in explaining why glacial periods tend to end so abruptly, much more abruptly than they begin. The factors in question include the ice-albedo effect, and the release of carbon dioxide from a warming sea and land. While we may not yet have the models to predict, with any degree of conviction, the probability or timing of the occurrence of each tipping point in the cascade, the general idea of tipping cascades is antecedently plausible, given their hypothesized role in periods where we know rapid warming has occurred.

Permafrost

Steffen's account is not a model of climate change, but a suggestive framework that might help direct the development of models. Perhaps more worrying than frameworks or models is what is currently happening on the ground. On Steffen's framework, melting of the Greenland Ice Sheet, West Antarctic Ice Sheet, Alpine glaciers, and summer Arctic ice are Stage 1 phenomena. Changes in ocean current, and resulting die-off in vegetation, are Stage 2 phenomena. The melting of permafrost is a Stage 3 phenomenon. However, there is ample evidence of significant melting of permafrost already.

According to the National Snow and Ice Data Center, the world's permafrost is estimated to contain around 1400 gigatons of carbon.[31] A gigaton is a billion tons. We will be working a lot in gigatons in later chapters, believe me. In contrast, the atmosphere, at present, contains around 850 gigatons of carbon. On a business-as-usual scenario, it is thought that roughly 70% of permafrost will melt before 2100. The carbon stored in this newly thawed permafrost will be broken down by microbes which use it as fuel. These microbes release the digested carbon either as carbon dioxide or as methane. This does not mean that all the carbon present in permafrost will find its way there. Estimates vary. But even lower end estimates of 10% are rather worrying. The math is not difficult. Seventy percent of 10% of 1400 gigatons is 98 gigatons. Divided by the 80 years that remain before 2100, this amounts to 1.225 gigatons per year for the next 80 years, on average. That may not seem a lot. In 2019, the world as a whole produced almost 37 gigatons of carbon dioxide. Next to that, 1.225 gigatons may seem like small potatoes. However, carbon dioxide has an atomic weight of 44, since it contains one carbon atom with atomic weight of 12 and two oxygen atoms with atomic weight of 16 each. So 37 gigatons of carbon dioxide equates roughly with roughly 10

---

[31] National Snow and Ice Data Center, https://nsidc.org/cryosphere/frozenground/methane.html.

gigatons of carbon. Against this figure, 1.225 does not seem so insignificant. In fact, 1.225 gigatons is roughly the amount of carbon put into the atmosphere by the United States in 2017. The output of permafrost, that is, would be equal to the second largest carbon-producing nation on earth (China is first). And that is on a relatively low-end estimate that assumes only 10% of the carbon contained in permafrost becomes converted to carbon dioxide. This sort of GHG release is not accounted for on current IPCC models.

This, however, is not the worst-case scenario. When permafrost thaws, the organic matter it contains gets eaten and digested by microbes. These microbes will either produce carbon dioxide or methane, depending on circumstances. If there is oxygen available, then they will produce carbon dioxide. If there is no oxygen, they will produce methane. In swamps and wetlands, the microbes tend to make methane. And, in general, swamps and wetlands are precisely what permafrost becomes when it melts. As we have seen, methane is a much more efficient trapper of heat than carbon dioxide, since it is a more complex five-atom molecule, compared to the three-atom carbon dioxide. Thankfully, it doesn't remain in the atmosphere anywhere near as long as carbon dioxide, but while it is here it traps heat up to 72 times more efficiently than carbon dioxide. For the next 20 years or so at least, the amount of methane released by melting permafrost is going to be crucial to our attempt to rein in global temperature increases.

The worry is that the melting of permafrost—a Stage 3 phenomenon in Steffen's account—is already upon us. In November 2018, when temperatures at the North Pole should have been –25°C, they instead registered a positively balmy 1.2°C. Throughout 2018 and 2019, the Arctic continued to warm at twice the rate of the rest of the planet. Arctic air temperatures for the past 5 years (2014–2018) have exceeded all previous records since 1900.[32] According to the 2017 National Oceanic and Atmospheric Administration (NOAA) Artic Report Card, "the Arctic shows no sign of returning to a reliably frozen region." Moreover, the "Arctic environmental system has reached a 'new normal,' characterized by long-term losses in the extent and thickness of the sea ice cover, the extent and duration of the winter snow cover and the mass of ice in the Greenland Ice Sheet and Arctic glaciers, and warming sea surface and permafrost temperatures." The 2019 Report Card continues in the same vein. The Greenland Ice Sheet—a Stage 1 phenomenon—is losing

---

[32] National Oceanic and Atmospheric Administration, Arctic Program, https://www.arctic.noaa.gov/report-card.

267 billion metric tons of ice per year. Crucially, the Stage 3 phenomenon, thawing Artic permafrost, "could be releasing an estimated 300–600 million tons of net carbon per year to the atmosphere."[33]

Reason and Ignorance

Under a business-as-usual scenario, the possibilities for the earth's future temperature trajectory range from, on the one hand, >2°C–5°C (>2°C is, essentially, the bottom limit of any business-as-usual projection) to, on the other hand, runaway global warming and a hothouse earth. Of course, as the World Bank has pointed out, even 4°C will still be very bad. So we have to conclude that if we continue with business as usual, prospects for our future range from the very bad to the, quite literally—or as literal as one can be when invoking a mythical realm—hellish. Which of these is most likely? There's the rub: we don't really know.

Consider the Arctic. There is widespread melting of the sort not seen for millennia. This will result in decay of the carbon contained in the permafrost. This will result in the emission of both carbon dioxide and methane. We know this. But what we don't know easily outweighs what we do. For example, warmer arctic temperatures mean a longer growing season, and so Arctic plants—dwarf shrubs, grasses, mosses, and lichens—will have longer to sequester carbon from the atmosphere. How much will this offset the release of carbon dioxide and methane? We don't know. At the moment, because of its vegetation, the Arctic acts as a carbon sink—it absorbs more carbon than it emits. If enough permafrost carbon decays, however, this will change, and the Arctic will become a carbon source. When will this occur? And how much carbon will have to be emitted before it does? We don't know. When the permafrost carbon does decay, being broken down by microbes, how much of this will be released as carbon dioxide and how much as methane? Having an answer to all of these questions is hugely important for assessing the climatic impact of GHGs over the next couple of decades. But we do not, as yet, have definitive answers, because they depend on a variety of unknown factors. For example, the ratio of methane to carbon released by melting permafrost—rather important to know, given the difference in

---

[33] National Oceanic and Atmospheric Administration, Artic Program, https://www.arctic.noaa.gov/Report-Card/Report-Card-2019.

relative warming potentials—will depend, in part, on how much of the newly melted permafrost will turn into wetland, since ponds and lakes and swamps will tend to produce methane more than carbon dioxide. How much permafrost will turn into wetlands? We don't know.

The case for anthropogenic global warming is, of course, a case grounded in science, as it should be. But science, naturally enough, always plucks the low-lying fruit first. We know that carbon dioxide, methane, nitrous oxide, and other GHGs are made up of three or more molecules. We know that and why gas molecules made up of three or more atoms absorb infrared radiation. We know the atmospheric presence of these gases has been rapidly increasing in recent decades. We know this increase is anthropogenic. We know this should increase global temperatures. We know that global temperatures are indeed increasing. All this is easy, low-lying fruit.

The harder stuff comes next. And this is where our ignorance starts to reassert itself. How much will a loss of polar ice reflectance drive up temperatures? How much will ocean salinity be reduced as a result of the melting of Arctic and Antarctic Ice Sheets and glaciers? How much will a given drop in ocean salinity impact on the Atlantic Meriodonal Current and the wider ocean conveyor? How much of a change in weather patterns will be occasioned by a 20% drop in vigor of the Atlantic Meriodonal Current? Or a 25% drop? How much decaying permafrost carbon will be released as carbon dioxide and how much as methane? These are harder questions, to be found on the higher branches of scientific inquiry. And we simply do not have the answers. We do have guesses, and some of these guesses are educated ones, supported by data and inferences based on this. But what we don't have is anything in which we can put an inordinate amount of faith.

This is how it is with human knowledge more generally. We humans are never as clever as we think we are. Human knowledge is the light of a tiny candle surrounded by the vast, dark night of human ignorance. What we don't know will always dwarf what we do. And, so, we must resign ourselves to doing the best we can under what are, in effect, conditions of profound ignorance. To augment our ignorance, we might add another consideration. The current atmospheric changes we are bringing about are entirely unprecedented in the history of the planet. It is not the raw numbers that are the problem. The earth has undergone massive changes in the past. The real problem is the *rapidity* of change. Given enough time, there is some reason to suppose that the planet could adapt to the newer, higher levels of atmospheric carbon dioxide and other greenhouse gases. A hotter planet would result in greater levels of vegetation which, if we did not slash and burn our way through it, would act to sequester atmospheric carbon. The problem is that the changes we are causing the planet to undergo are, from the perspective of a planet, inconceivably rapid. Cycles of glaciation typically take tens of thousands of years to complete. Roughly. Even that, in geological terms, is pretty rapid. However, the changes wrought by us in the planet are occurring in centuries or even decades, not tens of thousands of years. We are changing the planet on a timescale far too quickly for it to have a chance of adapting. Nothing like this has ever happened before on earth. And no one knows how it is going to play out.

We are in uncharted water—new and rather murky territory. How do we think when we are in the dark? How do we reason under conditions of profound ignorance? In the dark, you might, not unreasonably, take *caution* to be your cardinal, guiding virtue. But what counts as caution here? In general, to be cautious is to play it safe, to accentuate potential risk over possible reward. The risk is runaway climate change that results in a hothouse earth, not too dissimilar from Venus, and the extinction of all life on the planet. Perhaps this will happen. But perhaps not. This risk, one might think, is so great, some might think, that it is simply not worth taking. We should act on the basis of a worst-case scenario. Let us do *whatever* we can to avert this possible result. This strategy has its attractions. But, as I shall try to show in the next chapter, it also carries huge risks. I shall try to convince you that it is not caution that we need here.

4

The Fire

Energy, Civilization, and Collapse

The Hulk Coaster in the Age of Coronavirus

Our recent pestilential circumstances have given us all an unwelcome crash course in cost-benefit analysis. The benefit in question is, of course, staying alive. For some with COVID-19, the probability of death, averaged over all age groups, seems to be, at the time of writing at least (June 1, 2020), between 0.5% and 1%. These may seem like good odds, but they are not to be sniffed at. Imagine you were about to get on my sons' favorite thrill ride, the Hulk Coaster. There are, as far as I recall, 9 rows of 4 in each car. That's 36 people. Just as you are about to get on, you are told that for every three cars that make it around the coaster, one person is going to get killed. Would you still get on? I wouldn't. Awesome ride that it is—and I mean awesome—you would have to be insane to do so. But one death in every three cars equates to slightly below a 1% probability of death. Moreover, I doubt that being told not to worry, the death rate is, in fact, only one person in every six cars—slightly less than a 0.5% chance of dying—would substantially change my assessment. The moral to draw, it seems, is that the relatively low probability of a certain outcome can often be outweighed by its utterly calamitous nature. It might be different if, for example, that one person in every three cars would develop a mild headache on completion of the ride. Then, the attractions of the Hulk Coaster might well win out. But when the possibility is such a grave one, this can often—and quite reasonably—outweigh the low probability of its occurrence. Sufficiently serious possible outcomes, one might think, must be defended against at all, or pretty much all, costs.

We know from our current COVID-19 difficulties that there are drawbacks to this idea. COVID-19 has killed a tragically large number of people. If you catch the virus, there is a nonnegligible chance you are going to die. Precisely how nonnegligible this chance is will depend, as we know,

World on Fire. Mark Rowlands, Oxford University Press. © Oxford University Press 2021.
DOI: 10.1093/oso/9780197541890.003.0004

on your age and comorbidities. But, for anyone, there is, as far as we can tell, always a chance. Since death is the most serious and irreversible of possibilities, there will always be a temptation to allow its seriousness to override its improbability. We are all familiar with the consequences of this temptation: economic devastation and resulting loss of livelihoods, educational deprivation, loss of access to health care (ironically, just when you might need it the most). That strange-looking mole you were going to have your dermatologist take a look at, just before your local hospital shut down all nonessential procedures, looks a lot worse now in July than it did in February. It may be that more people die as collateral damage of COVID-19 than ever die from the virus directly.[1]

With respect to climate change, there is a spectrum of possibilities generated by the business-as-usual assumption. These range from, at one end of the spectrum, a very unpleasant global increase of greater than 2°C by 2100 to, at the other end of the same spectrum, something very different— runaway global warming leading to a dead or dying planet. What are the relative probabilities of any point on the spectrum occurring? The short answer is that we don't really know with any degree of certainty. It is generally accepted that somewhere toward the 2°C–4°C end of the spectrum is much more probable than the corresponding runaway global warming alternative. But the possibilities we can assign to any point on the spectrum are constantly changing. At any given time, we don't really know, with any certainty, what they are.

Runaway global warming would mean the death of the planet. If we assume that the death of the planet, occasioned by runaway global warming, perhaps brought about by the kinds of tipping cascade identified by Steffen and colleagues, is a nonnegligible possibility, then one might be tempted to think that this is a possibility so calamitous that it must be guarded against at any and all costs. Every step, one might think, should be taken to make sure it does not transform from possibility to actuality. However, as with the corresponding temptation vis-à-vis COVID-19, the costs must also be considered. This chapter is about some of those costs. Chapters 5–7 will begin sketching what I shall argue is a better approach: a way of protecting the planet while minimizing costs.

---

[1] See, for example, Bhavya Dore, "Covid-19: Collateral Damage of Lockdown in India," *British Medical Journal* 369 (2020): m1711, for an account of the kind of collateral damage in India occasioned by the lockdown.

An Exchange for Fire

A way of trying to sift through the falsities, fallacies, and fabrications that accompany discussions of climate change involves, fundamentally, the old philosopher's trick of *seeing the world anew*—that is, seeing the world for what it really is, and the deep principles and overarching concepts that underlie, drive, and unify it. We must move from the realm of what merely *is* to that of what *must be*. Of course, different people will have different ideas of what, precisely, these deep principles and overarching concepts are. The version, or vision, I am going to recommend can be traced back almost to the very birth of Western philosophy. The overarching concept is that of *energy*. And the deep principles are the first two *laws of thermodynamics*.

My favorite pre-Socratic philosopher, Heraclitus of Ephesus, once remarked: "This world-order (*kosmos*), the same of all, no god nor man did create, but it ever was and is and will be an ever-living fire, kindling in measures and being quenched in measures."[2] This was, perhaps, the earliest statement of what we now call the principle of *mass-energy equivalence*: anything having mass has an equivalent amount of energy, and vice versa, with the relation between the two enshrined in Einstein's equation, $E = mc^2$. Heraclitus put the matter thus: "All things are an exchange for fire, and fire for all things, as goods for gold and gold for goods."[3] We generally think of the world as a collection of things—of objects of varying sorts. Heraclitus's realization was that these things are really energy—or, as he put it, *fire*. Think of the world—cosmos—not as a collection of things but, like Heraclitus, as a body of energy. If we do this, I shall try to convince you, the future possible trajectories of the earth in the Anthropocene, and the human society that depends on it, will become much clearer. That is a subject of later chapters. What will also become clearer is the drawback of always privileging the severity of the consequences of business as usual over the probability of those consequences happening. That is what this current chapter is about.

Energy is the common currency of the universe. There are two laws that, for my purposes at least, are particularly important in explaining how this system of currency works. First, energy—currency—can be neither created nor destroyed, merely converted from one form into another. This is the first law of thermodynamics. Second, in an isolated system—a system

[2] DK22B30, trans. John Burnet, *Early Greek Philosophy* (London: A. & C. Black, 1892), p. 135.
[3] DK22B90, trans. John Burnet, *Early Greek Philosophy* (London: A. & C. Black, 1892), p. 135.

energetically cut off from its surroundings—entropy always increases.[4] Entropy is disorder. Thus, this second law of thermodynamics tells us that an isolated system will always become more disordered.

The most familiar application of these two currency laws is to biological systems. Like all living things, you and I are complex structures. In order to continue existing, we need to get energy from somewhere. Happily, we are not energetically isolated from our surroundings, and we get this energy from the things we eat, breaking them into their constituent components (molecules) and appropriating the energy that this breakdown releases. If we do not do this, we die, and it is then us that breaks down into our constituent components. This breakdown follows a distinct path from more complexity to less complexity—equivalently, from less to more disorder. This trajectory is entropy. In creatures like us, the input of energy we get through eating keeps the ravages of entropy—disorder—away, at least for a while. Plants obtain their energy differently, of course. But the same general principles apply.

While the laws of thermodynamics are most obviously applicable to living things—biological organisms—their application is, in fact, utterly ubiquitous. There is one complex structure, however, that provides the principal focus for the concerns of this chapter. It is not biological organisms, or the natural environment, or the ecosystems that make it up, more generally. Rather, it is society, human or otherwise.

Energy and the Evolution of Social Complexity

The application of the laws of thermodynamics to society may, initially, seem odd. But societies are complex structures, and as such the laws of thermodynamics will apply to them also. To remain in existence, any society needs to take in energy. Let us imagine the story of an initially simple society, one that gradually evolves into something more complex precisely through its attempts to acquire energy. Winters are harsh, and without fire—kindling in measures and being quenched in measures—the members of this society would freeze to death, and the society would cease to exist. Happily, however,

---

[4] There are two further laws of thermodynamics. According to the third law, the entropy of a system approaches a constant value as the temperature approaches absolute zero. According to the Zeroth law, if two thermodynamic systems are each in thermal equilibrium with a third one, then they are in thermal equilibrium with each other. These laws have been relegated to a footnote since they are not important for the purposes of this book.

this society has recently discovered a useful cache of stored energy, in the form of a large seam of coal located some way beneath the earth. Having largely used up other, easier-to-obtain, supplies of stored solar energy—that is, they have cut down all the trees in their area—this society now becomes a society of coal miners. They begin digging.

In time, of course, the same thing happens to the coal as happened to the trees. Initially, like the trees, the coal is easy to acquire. But the miners eventually use up all the easy-to-get-at coal, and now must begin to dig deeper. At the beginning, even the deeper seams are relatively easy to access. That, however, soon changes, and they must start digging further into the rock to get at the more recondite seams. At this point, perhaps, an important truth—one of the fundamental laws of currency—dawns on them. The value of coal is all about the *EROI*. It's all about the *energy returned on energy invested.*[5] Return on investment is the fundamental equation that will determine the fate of this society. Coal has value to the extent that the energy it contains can be used for important purposes. In this primitive stage of the society's development, this might amount to little more than keeping its members warm enough to survive, but further uses will emerge in time. However, what detracts from this value is the energy you must expend in order to extract the coal. The less energy you need to expend in extracting the coal, then the more value that coal will have for you.

The miners eventually come to realize that they might make certain changes to their mining practices that make coal extraction more efficient— that is, requiring less overall energy investment. Initially, everyone simply went to work digging the coal, and hauling it out of the mine shaft. That might not have been a bad strategy when the coal was plentiful and located near the surface. But as the diggers had to insert themselves further and further into progressively more difficult-to-access seams, where getting themselves in and out of position to cut coal was becoming more difficult, it became clear that a new strategy was called for. The miners, therefore, divided themselves up into two groups with different functions. On the one hand, there were the *colliers*, the cutters of coal. On the other, there were the *hauliers*, those responsible for hauling the coal to the surface. The simple coal-mining society has now become more *complex*. Previously it had one group of people doing

---

[5] The concept of energy returned on energy invested, EROI (or, sometimes, EROEI) was first introduced by the systems ecologist Charles Hall. It is central to the arguments I shall develop vis-à-vis animals and climate change, and we shall revisit it many times during the course of this book.

the same things, but now it has two groups of people doing different things. This increase in complexity is the attempt to solve a problem—the problem of how to extract coal most efficiently. And this problem is ultimately a problem of energy, specifically a problem of EROI. This is evident from what the word "efficient" means in this context: how to extract coal with the minimum expenditure of energy, so that the ratio of energy returned to energy invested will be as high as possible. Thus, the increase in complexity has been driven by the necessity of maximizing the EROI of the coal that is being mined.

As time goes by, and the available coal becomes more and more difficult to obtain, further changes might be implemented. Carrying coal out by hand, or in a sack, is a very energetically inefficient method. A mine cart, or tramcar, would be much better, allowing to you to carry greater quantities of stored energy to the surface while expending less energy in doing so. Then, however, you need people to build the tramcar and the rails. The society again becomes more complex. Moreover, it might soon be realized that the human ability to haul tramcars is paltry compared to that of a horse, and so pit ponies arrive on the scene. But then you will need further job functions. The ponies will need shoes, and so you need farriers. You might need someone whose function is to keep the pony healthy, during its miserable life down below. And, thus, is born the proto-veterinarian. In this way, the coal-mining society becomes progressively more and more complex.

This additional complexity is not because of an increase in the size of the population. There needn't be more people for a society to become more complex, and the simple addition of people will not, by itself, make a society more complex. A society of five, comprising a collier, a haulier, a tramcar builder, a farrier, and a veterinarian, is more complex than a society of a hundred colliers. It is more complex because it has more identifiable groups with distinct functions. The complexity of a society is a function of two things: the number of its component parts (classes, professions, and the like) and the variety of relations between these component parts.

The increase in complexity of this coal-mining society has been driven by an overarching imperative: to acquire the maximum amount of energy with the minimum expenditure of energy. We might call this the *EROI-c imperative*. When sources of energy are plentiful and easy to obtain, satisfying the EROI-c imperative is not really a problem. But when sources of energy become more recondite, it does become a problem. The increasing complexity of human (and, indeed, some nonhuman) societies is the attempt to solve this problem.

The Fall of the Roman Empire

The foregoing story is simply an illustration of an analysis of the development, and eventual collapse, of complex societies presented by Joseph Tainter, in his seminal work, *The Collapse of Complex Societies*.[6] The collapse of societies, empires, and civilizations is a well-documented phenomenon. However, there is little agreement on the causes of such collapse. Gibbon, in the first systematic study of the fall of the Roman Empire, attributed this to the decline of traditional values.[7] However, there are, reportedly, over 200 different accounts of why the Roman Empire fell.[8] And, as Tainter has noted, explanations of the collapse of other civilizations exhibit the same kind of variety.[9] These explanations divide into two: multifactor or single factor. Multifactor explanations attribute collapse to several different factors acting at once. For example, Cline argued that the collapse of Late Bronze Age Mediterranean civilization could be explained by way of a combination of negative factors, including climate change, earthquakes, and foreign invasions.[10] Tainter, however, has defended a compelling single-factor model of societal collapse. He argued that societal collapse—which he understands as a reduction of societal complexity—can invariably be traced to one thing: *diminishing returns*. Once the EROI—the ratio of energy return on energy invested—falls below a certain point, societal collapse is an inevitable result.

Today, solar power is regarded as a key form of renewable energy. In fact, however, all energy found on earth is ultimately solar in the sense that it derives from the energy of the sun. The only difference lies in whether it is stored or not. Whether we burn oil, coal, gas, or wood, this is all energy that has been provided by the sun, energy that nourished the trees and animals that eventually turned into these, as we put it, *resources*. The Roman Empire was, in this sense, a solar power. The Empire was built on a policy of loot and pillage. Wherever and whenever it expanded, it would acquire new sources of energy that ultimately derived from the sun, solar energy incarnated in animals, trees, crops, and people.

[6] Joseph Tainter, *The Collapse of Complex Societies* (New York: Cambridge University Press, 1988).

[7] E. Gibbon, *History of the Decline and Fall of the Roman Empire* (London: Strahan & Cadell, 1776).

[8] A. Demandt, *Der Falls Rom* (Zurich: Beck'sche, 1984).

[9] J. Tainter, "Collapse, Sustainability, and the Environment: How Authors Choose to Fail or Succeed," *Reviews in Anthropology* 37 (2008): 342–371.

[10] E. H. Cline, *1177 B.C.: The Year Civilization Collapsed* (Princeton, NJ: Princeton University Press, 2014).

Minerals are, also, a form of energy—or can be used as such. Minerals often constitute what we might call a *proxy*, as opposed to *direct*, source of energy. A tree, for example, is a direct source of energy. You can burn it and release this energy, which can then keep you warm, drive the blacksmith's forge, and so on. Minerals are a proxy source of energy: with them you can make coins, and with these coins you can, via a largely arbitrary series of human conventions that imbues these with value, get people to work—that is, expend energy—for you. Or you can buy things, such as coal or timber, that you can use for energy and for other things that people have had to invest energy in acquiring or making (swords, ploughshares, and so on).

Whether direct or proxy, this energy upon which the Romans had managed to get their hands, had been built up over thousands or millions of years, and the Romans acquired and used it in a matter of decades. As Tainter points out, the problem they then faced was clear: they had an empire to run and now had to maintain it using present-day rather than stored energy. The initial response to this problem was obvious: further expansion and the acquisition of more stored energy. But this merely postponed the problem. Eventually, even if they had managed to conquer the whole world, the Romans would run out of stored energy, and present-day energy would be all that remained. Their response to this problem was an impressively complex society designed, Tainter argues, with the aim of acquiring energy with maximum efficiency. It is, as in the case of our imagined society of coal miners, the necessity of acquiring energy with the minimal expenditure of energy—keeping the EROI as high as possible—that drives complexity.

The eventual collapse of the Roman Empire, Tainter argues, in a detailed and compelling analysis, arises from the ultimate inability to keep the EROI of their energy sources high enough. At the heart of the problem is the fact that complexity has enormous costs as well as benefits. Brief reflection on the laws of thermodynamics we encountered earlier quickly conforms this. The more complex a structure—whether a living organism or a society—the more energy is required to maintain its complexity. Complexity is, therefore, an ultimately self-defeating strategy. Societies evolve complexity in order to solve problems the world has set them. The fundamental, baseline, problem is one of acquiring energy at a suitable EROI value. Given this is so, the strategy of evolving complexity in order to acquire energy has a definite, if unpredictable, shelf life. More complexity requires more energy, and eventually the sources of energy are going to run out. Therefore, the costs of complexity will, eventually, come to outweigh the advantages. That is why complex societies—such as the Roman Empire—collapse.

While I find Tainter's account of why complex societies collapse to be convincing, it is not universally accepted. When developing any argument, it is advisable to reduce one's required assumptions to the minimum number possible that are compatible with the success—validity, soundness—of the argument. In this spirit, I should point out that my arguments do not depend on Tainter's specific explanation of societal collapse as involving ever-increasing complexity with ever-diminishing returns. What is essential to the argument I am going to develop—in this chapter and the next—are two claims. First, as a complex structure, any society is going to require a net input of energy in order to remain in existence. Second, the long-term prospects of that society depend on the EROI of its energy sources. Once these drop below a certain point, the society will become nonviable.

The Net Energy Cliff

In most essential respects, we are like the Romans. Theirs was an empire of people, land, animals, crops, and minerals. Ours is an empire of coal, oil, and gas. We have used energy from the sun, stored in the form of fossil fuels, that took millions of years to create. And we have used much of it in a matter of generations. The problem of climate change that we now face is a byproduct of our fossil fuel empire. And although the climate is changing apace, we nevertheless have large, complex societies to maintain—societies that have become large and complex precisely in the service of efficient acquisition of stored energy. Accessing stored energy will become increasingly difficult as the reserves dwindle. And using this stored energy will become increasingly unjustifiable as their greenhouse gas (GHG) effluents further saturate the atmosphere. Eventually, at some point, we will have to make do with present-day energy. We all know how things turned out for the Roman Empire. Becoming ever larger and more complex was not enough to save it from collapse. According to Tainter's analysis, the reason for this is that the energy-through-complexity strategy is subject to the law of ever-diminishing returns, an ever-diminishing ratio of energy returned to energy invested. Becoming more complex may allow you to acquire energy more efficiently. But this increased complexity does not come for free: it needs more energy to sustain it, which in turn requires more complexity, which requires more energy. The cycle cannot go on forever, and when it breaks down, the result is collapse. Whether or not this account is correct—as I mentioned, my argument does not require this assumption—our key question is an obvious one: Can we avoid what happened to the Romans, and perhaps all other civilizations?

It is tempting to think that the position we are in is crucially different from that of the Romans. We have something they did not: *modern technology*. And with it, we can harness an unlimited supply of renewable energy, from the sun, the wind, and the tides. Or, if we wanted to go in a somewhat different direction, we might use nuclear fission. The Romans collapsed, ultimately, because of their reliance on historical, stored energy. But we have the technology available to build wind farms, solar farms, tidal barrages, and nuclear power plants. Would this not allow our society to endure? Answering this question requires that we answer two further questions. First, we need to work out the kind of EROI we are going to need in order to maintain our society in its current form, or something recognizably similar to its current form. Second—and this will be the subject of the next chapter—we need to work out which forms of energy, if any, can supply this required EROI.

The question of what EROI is required for maintenance of a society recognizably similar to our own has been the subject of more than a little study in recent years. Charles Hall, one of the earliest researchers to address this question, used to think that an EROI of 3 would be enough to sustain a society recognizably similar to our current one. However, he has recently changed his assessment and now argues that we may, in fact, need an EROI of somewhere in the region of, at a minimum, 12–14. Thus, Lambert, Hall, and colleagues identify a pyramid of societal energy needs.[11] From the bottom up, it runs as follows: find energy, extract energy, build and maintain transportation systems, grow food, support families, educate the young, provide health care, and support the arts. An EROI of 1 gives you nothing. You have expended as much energy locating the resource—let us suppose it is oil—as you would get from it. According to Lambert and Hall's calculation, if you have an EROI of 1.1:1, you can extract the oil from the ground. An EROI of 1.2:1 will allow you to refine it. An EROI of 1.3:1 will allow you to move the oil. But if you want to distribute it, via a truck, then you will need at least 3:1. But if you want to transport other things, like grain, you will need an EROI of 5:1. That includes depreciation for the truck (trucks will eventually need

[11] J. Lambert, C. Hall, S. Balogh, A. Gupta, and M. Arnold, "Energy, EROI, and Quality of Life," *Energy Policy* 64 (2014): 153–167.

replacing). But if you want to include the depreciation for the truck driver, the oil worker, and the farmer (they will all need food, water, shelter, and eventual replacement via theirs or someone else's children), then you've got to support them and their families. And for this you will need an EROI of 7:1. If you want an education system, you'll need an EROI of between 8:1 and 9:1. Health care will increase that to 10–11:1. And the arts will take that up to 14:1. Thus, as an answer to the question of what EROI is required to maintain a society recognizably similar to our own, Lambert and Hall arrive at the figure of 12–14.

David Murphy, in a major survey article commissioned by the *Royal Society*, also arrives at a not-too-dissimilar figure: an EROI of at least 11 (or, he thinks, probably higher) is required to sustain a society recognizably similar to ours.[12] Fizaine and Court also, independently, identify the figure of 11 as the minimum required threshold of societal EROI.[13] In the next section, we shall look at some wrinkles in this idea which suggest that even 11–14 might be far from ideal.

The idea driving these calculations of a minimum required EROI is the concept of a *net energy cliff*, a concept originally devised by the geochemist Euan Mearns. The concept of *net energy* is closely related to EROI. EROI is the energy returned *divided* by the energy invested. Net energy gain is energy returned *minus* energy invested. Net energy measures the overall *quantity* of energy you obtain from a given source, whereas EROI measures the *ratio* of energy acquired versus invested. If you get 10 kilojoules of energy out of a given source—a certain amount of coal, for example—and you only had to expend 1 kilojoule of energy extracting it, then the EROI is 10 divided by 1, whereas the net energy gain is 10 minus 1. The EROI is 10, but the net energy gain is 9. The actual formula relating the two concepts, EROI and net energy gain, is as follows:

Suppose that the net energy of a given source—coal, oil, or whatever—decreases because it becomes difficult to locate or extract. Then, if we want to maintain the same output of energy, we will have to allocate more resources—equipment, people, money, and so on—to locate and extract the same quantity of energy. But you have to maintain—that is, feed, clothe, house, and so on—all the people who are doing this locating and extracting, and offset for increased depreciation of their tools and equipment. Therefore,

[12] David Murphy, "The Implications of the Declining Energy Return on the Investment of Oil Production," *Philosophical Transactions of the Royal Society A* 372 (January 2014), 1–19.

[13] F. Fizaine and V. Court, "Energy Expenditure, Economic Growth and the Minimum EROI of Society," *Energy Policy* 95 (2016): 172–186.

more quantities of energy must be inputted simply to extract the same quantity of energy as before. But this energy does not come from nowhere: it needs to be obtained from sources that are increasingly difficult to locate and extract. The result is a vicious circle. When the net energy of a source is around 100, as it was for coal and oil in the good old 1930s, Great Depression aside, comparatively few people and resources have to be allocated to finding and extracting this source. Conversely, if the net energy of a source were 1, every single person and resource in a society would have to be allocated to finding and extracting it. This society would be running just to stay in place. Between a net energy value of 100 and around 20, there is only a shallow drop-off; slightly more energy must be invested to extract the same amount of energy. However, once we get below a net energy of around 15–20 that starts to change. At a net energy value of 10, roughly 10% of our total amount of energy must be used to extract the same amount of energy, leaving far less energy available to be diverted elsewhere. Somewhere around here, or just before, is the tipping point. After this point, so many people and things have to be dedicated to locating and extracting energy that their energy needs outweigh the energy they are able to extract. The result is that when we reach a net energy value of 10, or perhaps slightly above, we begin to fall off a net energy cliff. The edge of the net energy cliff is passed when we allot so much energy to the task of acquiring energy that there is not enough energy left over to maintain or fulfill the other needs of our complex society. As long as we keep the denominator as 1, net energy is EROI minus 1. Thus, once our energy sources reach a net energy value of 10, or EROI of 11, or somewhere in this general vicinity, we should expect our society to become unsustainable in its present form.

Energy and Liberal Democracy

Based on these sorts of considerations, answers to the question of the minimum EROI required to sustain a society recognizably similar to ours, therefore, have tended to coalesce around the 11–14 mark. This is not an exact science, however, and a lot turns on what is meant by a society "recognizably similar to ours." There are at least three drawbacks involved in framing the issue in this way.

The first is the inherent vagueness of the concept of recognizable similarity as employed in this context. Is, for example, the society of 1960s "recognizably similar" to ours? It is not clear there is a correct answer to this question. We might say that it is recognizably similar, but this would mask some rather notable differences. When I was a young kid growing up in Wales, I would get to take a proper bath once a week—whether I needed it or not. (The rest of the time it was a bowl and a sponge.) My younger brother would have to share the bath water, so as not to waste it. For food, my mother used to make a stew, largely potatoes and carrots, with a little lump of lamb that she had got from a can. We used to dip a lot of bread in it to fill up, and this would last us most of the week. We weren't particularly poor. Not at all—this was the norm. We had a car, for example. Not everyone did. Only one, though. Is the society of my childhood "recognizably similar" to the society of today? You might say that it is, but if so, we would have to accept that some rather large differences reside under the rubric "recognizably similar." Or, in the 1970s, when a convergence of various energy crises began to bite, but the immediate cause being the coalminers' strike, the UK government implemented a 3-day working week to preserve available energy supplies. I remember thoroughly enjoying the days off school. Is this 1970s society recognizably similar to ours? As we shall see, in this century, the heyday of our fossil fuel empire, many of us have grown up in the warm embrace of EROIs typically ranging between 30 and 100. It is far from a given that an EROI of, say, 14 would keep us in anything like the lifestyle to which we have become accustomed. It may well be more like the United Kingdom of the 1960s and 1970s than the Western, developed world of 2020.

The second drawback in framing the issue in terms of the idea of a society recognizably similar to our own is that it is parochial. It is, of course, not just the energy needs of our society that are important but also those of developing societies. Framing the issue as a "What EROI do we need to sustain a society recognizably similar to ours?" question masks this fact. Some of us may be perturbed by a return to the one-bath-a-week scenario I mentioned in the previous paragraph, but to those in at least some developing societies this may seem entirely reasonable. In other words, the question "What EROI do we need to maintain a society recognizably similar to ours?" can mask a significant divergence in the notion of the "we" and "ours." It is not clear why the "we" and "ours" of the developed world, which might be distressed by an EROI in the vicinity of 10, should be privileged over the "we" and "ours" of the developing world who might regard this as an upgrade.

The third drawback is that the question might cause us to think of the issue in all-or-nothing terms. A certain threshold is reached—look!—the society is recognizably similar to ours, and whatever EROI we needed to get us here, that is the EROI we need. It is far more realistic, and therefore, useful to think about this issue in terms of a spectrum of cases—a range of societies which, due to progressively declining EROI, deviate progressively from ours—as opposed to a dichotomy according to which a society is either recognizably similar to ours or it is not. This latter point is particularly important because there is much more at stake than simply our levels of comfort or convenience. The nature of our society, and its degree of democratic openness, are also in question.

The consequences of a low EROI have, historically, been rather grim, and there is no obvious reason for thinking they will be hugely different today. In the developed world at least, we have grown up in a historically almost un-precedented era—an era of Empire, the empire of coal, gas, and petroleum. Not since empires of conquest, the Roman Empire, for example, and perhaps not even then, have people had it so good in terms of EROI.[14] And reflections on the fate of the Roman Empire are hardly comforting. The effects of a diminishing EROI were initially masked, in effect, by a *devaluation in proxy energy*. In the year 0 AD, the denarius was a 98% pure silver coin. As the years passed, it progressively came to contain less and less silver. By the year 269 AD, it contained a mere 1.9% silver.[15] This currency devaluation, how-ever, merely delayed the inevitable. The first results of this devaluation of the denarius were huge rates of inflation followed by a series of financial crises. There are reports of people having to wheel cartloads of denarii to the stores to buy even basic, essential, items.[16] Think Venezuela, today. As an attempt to cope with the financial crises, taxes were progressively raised—doubling and redoubling in the second and third centuries AD. As a result, people were unable to have families as large as had become customary, and the overall population fell, resulting in further pressure on tax revenues. In response, the Roman state became more and more authoritarian. Occupations were frozen into hereditary roles. If your daddy was a cobbler, then so were you. If he was a soldier, that was your lot, too. Eventually, some people became so poor they, reportedly, had to sell their children into slavery because they couldn't afford to feed them. The end result was collapse.[17]

---

[14] When the Romans started their era of expansion, they immediately abolished taxation of them-selves. We didn't quite get that far—unless you are Amazon, of course.

[15] For discussion, see Tainter, *The Collapse of Complex Societies*, 132.

[16] Tainter, *The Collapse of Complex Societies*, 128–142.

[17] See Tainter, *The Collapse of Complex Societies*, 128–151 for detailed discussion.

I certainly hope that selling our children into slavery is not going to happen here. But while we may not—and, as I shall argue later, need not—fall this far, the beginnings of diminishing EROI, perhaps the harbingers of collapse, are already visible. What are you going to do when energy becomes scarce? You might—this would be a natural response—circle the wagons and protect what is yours. You might build walls so people cannot come in and take your remaining energy. You might exit (or Brexit) multinational institutions, such as the European Union, and retreat to your own Little England. You might find yourself electing regimes altogether more authoritarian than have seen the light of day in the last half century or so. At the same time, you might purposefully strip away your environmental regulations—just when you need them most. You give notice of withdrawal from the Paris Agreement. The sorts of carbon sequestration initiatives required to get a grip on GHG emissions are done away on the grounds they are "job-killing regulations." These are no accidents. They are all connected. They are symptoms of a falling EROI.

It is no accident that the birth of modern democracy coincided with a time of unparalleled access to high-quality sources of energy. It coincided with the Empire of coal and oil. It coincided with high EROI. When energy is plentiful, the good times can roll. *Laissez les bon temps rouler*, as they say in New Orleans. Or used to say, before Katrina. Like a drunken tourist on Bourbon Street during Mardi Gras, one keeps handing out money for beads, tips for the bartenders and waitresses, emoluments for the musicians and the hustlers. But when times get tight, your mood changes—veers in an altogether more severe, authoritarian direction, as happened in the ancient Roman Empire. Less New Orleans, more Stalingrad. As EROI decreases, it may be that the world will veer in an altogether more authoritarian direction, too. Indeed, all the indications are that it is already doing so.

There are certain accepted markers of successful liberal democracies. Lambert, Hall, and colleagues have argued that many of these are strongly correlated with overall societal EROI.[18] These include the markers captured in the Human Development Index (HDI): a statistical composite of life expectancy at birth, years of education, and gross national income. Lambert,

[18] J. Lambert, C. Hall, S. Balogh, A. Gupta, and M. Arnold, "Energy, EROI and Quality of Life," *Energy Policy* 64 (2014): 153–167. We shall look more closely at the notion of societal EROI in the next chapter. The general idea is suitable for present purposes. You can think of it as the weighted, aggregated EROI produced by all energy sources employed by a society. This, if a society gets 60% of its EROI from coal, with an EROI of 30:1 and 40% from oil with an EROI of 20:1, then the societal EROI is (30 × 60%) + (20 × 40%), yielding a societal EROI of 26:1.

Hall, and colleagues have shown that nations with a societal EROI of less than 15:1 have HDI values below 0.7, whereas those with an HDI above 0.7 have societal EROIs of above 15:1. There are similarly strong correlations between societal EROI and other markers of successful liberal democracies. In countries with societal EROI of less than 15:1, typically more than 12% of children are underweight. However, in countries with societal EROI of above 20:1, less than 5% of children are underweight. The level of societal EROI also strongly tracks the level of per capita public health expenditure. Countries with higher societal EROI also have high female literacy rates, whereas for countries with low societal EROI, the rate of female literacy is far more mixed. The level of societal EROI strongly tracks levels of gender inequality: lower societal EROI almost always translates into higher levels of gender inequality. Finally, there is a similarly strong tracking relation between societal EROI and access to potable water.

Once upon a time, people used to think history had a direction. This conceit was not the sole preserve of a political Left inspired by Hegel and Marx. The Right also believed in it, thinking that its preferred form of economic liberalism was the "end of history."[19] Both parties believed you could be on the *right* or *wrong* side of history, although they disagreed on precisely what these sides were. These are all dangerous conceits because they conceal the peril we are now in. History does not have a direction, nor an end, and there is no right or wrong side of history. History is a long time. And for the most part it is quite unlike what we have become used to in the past 200 years or so. In the long time that is history, empires are just blips, the exception rather than the norm. Our Empire of fossil fuel is, statistically, merely an aberration, as was the Roman Empire era of expansion by conquest. High EROIs, of the sort with which we have all grown up, are the exception and not the norm.

At the beginning of this chapter, I talked about the downside of always privileging the potential severity of the consequences of business as usual over the probability of those consequences obtaining. This, then, is that downside. If there is one thing that our use of fossil fuels has supplied for us over the last 150 years or so, it is consistently high EROIs. It is likely that these positively luxuriant EROI levels have played a significant role in securing many of the markers of successful liberal democracies, as they are usually

---

[19] Most famously, and perhaps most disastrously, the end of history—in the form of Western liberal democracy—championed by Francis Fukuyama in his book, *The End of History and the Last Man* (New York: Free Press, 1992).

understood. If we want to simply jettison our use of fossil fuels—as the option of privileging the severity of consequences over the probability these consequences might have us to—then we had better make sure we have some alternatives in place that preserve our historically high EROIs, or something close to them.

If we do not, the danger is that we will lose our society in its current form, and the society that comes to replace it will, in a variety of ways, be far less appealing. Nevertheless, as we saw in Chapter 3, we cannot afford to ignore climate change—this baleful byproduct of our fossil fuel empire. We cannot do this because the consequences of business as usual will range from very nasty to existential threat. We cannot afford to ignore it, any more than the Romans could afford to ignore the repeated, and increasingly bold, incursions from the Germanic tribes, the Goths and Visigoths. The Romans were ultimately unable to do what they needed to do. The question is: can we? Can we do what needs to be done while maintaining in existence a society recognizably similar to our own? A society that is not only comfortable and convenient but also, and more importantly, open and democratic? Can we succeed where the Romans failed? Thinking of these issues in terms of the overarching concept of energy, and the guiding principles of EROI and net energy gain, has allowed us to at least understand what we need to do.

First, at a minimum, it seems likely that if we want to maintain a society recognizably similar to ours—replete with education, health care, and the arts—we must maintain a societal EROI of at least 11–14. However, even this may not be enough. The figures of Lambert et al. suggest that if we want to maintain more subtle markers of successful liberal democracies, we may need to maintain a higher societal EROI than this. If Lambert et al. are correct, societal EROIs of less than 25:1 tend to result in societies with poor or moderate quality of life—as measured in terms of markers such as HDI index, child malnutrition, public health expenditure, female literacy, and gender equality. Somewhere between 20 and 30:1, they argue is the transition point, where a society transforms into one with a good quality of life, thus measured. After we reach 30:1, little difference is engendered through further augmentation of EROI. Therefore, it seems that while a societal EROI of 11–14:1 is a minimum target, we ideally need to be raising our sights considerably higher than that. And, crucially, we need to do all of this, maintain societal EROIs of between 11 and 25+ while dramatically cutting our GHG emissions.

Can we do it? Is it even possible? I shall argue that it is and we can. It is not going to be easy, and it is not going to be palatable—especially to those with certain palates. But we can do it. How, precisely, is the topic of Chapters 6 and 7. But before we get to that, we are going to address a common idea—the idea that technology will save us. Technology is crucially important, of course. But it is not going to save us, not on its own. That is what the next chapter is about.

5

Salvation Technologies?

Carbon Capture and Sequestration

The Roman Empire eventually collapsed. Every other empire ultimately suffered the same fate. But, it is tempting to suppose, our empire of oil, coal, and gas has an advantage afforded to no other: modern technology. Perhaps our *technology will save us*. If the arguments of Chapter 4 are correct, we now have a relatively firm idea of what technology is going to have to do if it is, indeed, to save us from the climate change we have instigated without occasioning a significant societal downside. It must be able to drastically reduce greenhouse gas (GHG) emissions—ideally to zero, but we can welcome any reduction that is sufficiently dramatic—while keeping the energy returned on energy invested (EROI) of our energy sources in, at the very least, the 11–14 range. As we have seen, if Lambert, Hall, and colleagues are correct, we may have to aim considerably higher than this: somewhere in the 20–30 range if we are to hold on to a society that exhibits the common markers of modern liberal democracies. The question is: can any of our technologies achieve these two goals?

The first question to ask, in our attempt to drastically reduce or even eliminate GHG emissions, is this: do we keep or do we ditch fossil fuels? If we decide to ditch them, then we have essentially two technological options: *nuclear* or *renewable*. We will look at these options later on. If we decide to keep fossil fuels, then our assumption is that we can, in some way, render them *clean*. That is, we are placing our eggs in the *carbon capture and sequestration* (CCS) basket.

To *sequester* something is to isolate it or hide it away. To sequester atmospheric carbon—or carbon dioxide, the term is typically applied to both—is, accordingly, to take it out of the atmosphere and put it somewhere where it will do less harm: ideally, somewhere where it can't find its way back into the atmosphere for a very long time, if ever. Not all ways of sequestering carbon need be technological. Trees are very good at sequestering carbon, for example. So, too, are marine micro-animals such as phytoplankton. I shall

World on Fire. Mark Rowlands, Oxford University Press. © Oxford University Press 2021.
DOI: 10.1093/oso/9780197541890.003.0005

consider natural forms of sequestration in Chapter 7. My focus, here, however is on technological means of carbon sequestration.

The technological process of CCS works by, first, separating carbon dioxide from other gases contained in industrial emissions. Then, the carbon dioxide is liquefied, either through compression or cooling, or both, depending on the specific type of CCS apparatus employed. Finally, it is transported to a location that separates it from the atmosphere. Suitable storage areas, or areas that have been regarded as suitable, include natural geological formations such as *deep saline formations*. These consist in sedimentary rocks that contain pore spaces that are filled with water containing high concentrations of dissolved salts. They also include human-made spaces such as depleted oil and gas reservoirs. More controversially, the deep ocean has also been suggested as a suitable receptacle of carbon dioxide, a proposal that has now generally been rejected on the grounds that it will result in ocean acidification. CCS typically involves capture of carbon dioxide directly at the source of emission, from power plants and other types of industrial plants. In these cases, it works by stopping carbon dioxide emissions before they reach the atmosphere. However, CCS can also include attempts to remove carbon dioxide from atmosphere, by means of scrubbing towers or "artificial trees."

The problems that CCS technology must overcome are well understood.[1] They stem, ultimately, from some rather unfortunate raw numbers. Suppose you own a coal-fired power plant. Certain newly imposed environmental regulations aimed at cutting GHGs—courtesy of your country's signing up to the Paris Agreement, perhaps—require you to invest in CCS technology. The first problem you are going to face is that the amount of coal your plant burns is a quite a bit less than the amount of carbon dioxide it produces. Carbon dioxide is the result of a chemical reaction between the carbon in the coal and the oxygen in the atmosphere. Carbon, as we saw in Chapter 3, has an atomic weight of (roughly) 12. Oxygen has an atomic weight of (equally roughly) 16. Each carbon dioxide (CO_2) molecule is a combination of two oxygen atoms and one carbon atom, and so has an atomic weight of 44. Therefore, roughly 3.6 times as much carbon dioxide is produced for each unit of carbon burned. Coal is around 51% carbon. Therefore, roughly 1.8 times as much carbon dioxide is produced for every unit of coal that is burned in your plant.[2] This

---

[1] For a lucid discussion, see Michael Barnard, https://cleantechnica.com/2016/01/19/carbon-capture-expensive-physics/.

[2] At this point, perhaps, you might wish you had invested in a natural gas power plant instead, but this wouldn't help very much—methane produces roughly 2.75 times the weight of carbon dioxide per unit of methane burned.

discrepancy in size between source fuel (coal) and effluent (carbon dioxide) makes it likely that the mechanisms you are going to need for capturing and separating the carbon dioxide will have to be larger and, therefore, probably more expensive than the mechanisms used for burning the coal in the first place.

Moreover, these capture and separation mechanisms are not going to run themselves. Carbon dioxide is captured as a gas and, as such, is very diffuse. In order to be stored in reasonably sized receptacles, it must be turned into liquid form. This can be achieved either through compression or cooling, or a combination of each. But either option will require large amounts of energy, and this will further add to the expense. Any sensible CCS design will attempt to use the heat already generated by your plant for this purpose. Nevertheless, this heat will have to be appropriately directed to your capturing mechanisms, and it must be done so in the right amounts. So you will need more duct work, more processing, more fans, and more controls. All of this will entail further expense.

Capturing the gas is only the beginning of your problems. Now you have to work out what to do with it. After liquefaction, carbon dioxide must be stored onsite until it is ready for shipping. This carbon dioxide, however, will weigh around 1.8 times as much as the coal that was burned to produce it. Therefore, large pressurized and/or insulated vessels will be required to store the carbon dioxide. This adds further expense: it is far more expensive than the storage of coal, which can simply be piled up on the plant or factory floor.

The same issues of expense dog the distribution of the liquefied carbon dioxide. Distribution of liquefied carbon dioxide is, of course, much more challenging than the distribution of coal. Coal can be transported by train in open hopper cars. If you were to transport your carbon dioxide effluent by train, you would do so either using pressurized containers or pressurized containers that are also maintained at a very low temperature. The total number of train cars this would require would be greater than those required to bring in the coal in the first place.

Perhaps, you might decide, your carbon dioxide effluent is best transported via pipeline. The problem is that there aren't many of these pipelines around. In most countries these pipelines simply don't exist. There are several in the United States, but many more would be needed to be built if we are to rely on widespread CCS to mitigate our GHG emissions. This will add massive infrastructure costs to the expense of the CCS strategy.

Finally, the captured carbon dioxide must be safely stored. A suitable receptacle must be found that will insulate it from the atmosphere, ideally forever, but we might settle for a very long time. Here is where the current perversity of the CCS strategy manifests itself most completely. I mentioned that, unlike in most other countries, there are a few carbon dioxide pipelines in the United States. Almost all of these terminate at what are known as *enhanced oil recovery fields*. Pumping carbon dioxide into a played-out oil field increases the pressure underground and forces the remaining oil to flow toward the other end of the field where it can be pumped out. I kid you not. The end result of these massively expensive efforts to capture and separate, compress, store, and transport carbon dioxide is that we get our hands on more fossil fuels, which we can burn and produce yet more carbon dioxide.

Matters become even worse when we consider the nature of a played-out oil field. Such a field will be an area that contains dozens, hundreds, or even thousands of holes, both natural and human-made, in the form of natural faults and oil wells. The likelihood of carbon dioxide remaining safely sequestered here is minimal. Perhaps we could plug all the holes? Even if technically possible—which is a big if—this is going to add massively to the costs of CCS.

The culmination of these massive expensive attempts to sequester carbon dioxide, therefore, does not seem very different from business as usual. CCS, as we currently practice it, is less an attempt to sequester carbon dioxide and more an attempt to get our hands on yet more fossil fuels. That, of course, might seem like an inessential part of CCS. We could always eschew the played-out oil field part of the process and concentrate instead on sequestering the carbon dioxide in safer, more insulated, places, such as deep saline formations. We could. But getting difficult-to-obtain oil out of the ground was a crucial part of the commercial viability of CCS. We have seen the kinds of expense to which CCS will commit us: significant expenses that will attach to every step of the process, from capturing and separating, compressing, storing, transporting, and depositing carbon dioxide effluent. The lack of any financial offsetting from newly sourced oil would only exacerbate this problem. And it is the financially prohibitive nature of CCS that will be its ultimate undoing. If CCS does not work financially, then it is unlikely that it will work at all.[3]

---

[3] Even organizations that promote CCS acknowledge its financial costs. Estimates of the cost of CCS generally coalesce in the region of $120–$140 per ton of carbon dioxide. If correct, this has been estimated to add, roughly, an extra 17 to 20 cents to the cost per kilowatt hour (KWH) of a

The Declining EROIs of Fossil Fuels

You might be tempted to think: finances be damned! If this is what is required to reduce our GHG emissions, so be it. Any amount of money is worth the cost. But the financial considerations are simply reflections of a deeper, and more important, issue of EROI, or net energy—as we should expect since, as we saw in Chapter 4, money is a proxy form of energy. It takes energy to build and run CCS devices. It takes energy to build and run compression and liquefaction machinery, and storage receptacles for liquid carbon dioxide. It takes energy to build the pipelines or drive the trains that will transport the liquid carbon dioxide. All of this energy required to separate, capture, store, transport, and deposit carbon dioxide will inevitably erode the EROI of coal. The same is, of course, true of the other fossil fuels, which will require the same kinds of CCS technology if we are to mitigate their GHG contributions. This is a particularly vexing problem for the attempt to render fossil fuels clean. The reason is that the EROIs of these fuels are, already, steadily declining—for reasons entirely independent of the use of CCS technology—and have been doing so for some time.

If the EROI of coal and other fossil fuels were high enough, this problem would be surmountable. We could simply take the hit of a reduction in EROI occasioned by the CCS technology. The crucial question, then, is: what kind of EROI do fossil fuels now have? This is by no means an easy question to answer. It is well documented that the EROIs of fossil fuels have been declining, more or less consistently, over the past 80–100 years, some more consistently than others, largely due to our having used up to the easy-to-find-and-access sources and having to move on to more difficult ones. The more difficult a repository of coal, oil, or gas is to locate or access, the more energy will have to be expended in obtaining it, and the lower its resulting EROI will be. At one time EROIs of coal and oil were generally accepted to around 100:1, possibly even higher. Now fossil fuel EROIs are thought to average out—coal is the highest, with oil and gas lagging some way behind—somewhere in the region of 25–30:1. Until recently, however, the consensus was the EROIs of fossil fuels would remain at or above 25–30:1 for the foreseeable future—by which time, we would hope, renewables would be properly up and running. This

coal-fired power plant. Even 17 cents would place your coal plant deep into uncomfortably unprofitable territory. Other estimates—made by those less favorably disposed toward CCS—are often much higher, ranging as high as $3000 per ton of carbon dioxide. See https://www.huffpost.com/entry/the-strongest-case-against-nuclear-power-isnt-about_b_59ae014de4b0bef3378cdacf.

assumption, however, is looking increasingly optimistic. This is the result of a degree of conceptual variegation in the idea of EROI.

In their calculation of EROIs required to sustain a society recognizably similar to ours, or to preserve markers of liberal democracy such as a relatively high HDI rating, childhood health, female literacy, and so on. Lambert and Hall were working with what is known as *mine-mouth* or *well-head* EROI. However, this is not the only available concept of EROI. Following Murphy and Hall, it is useful to subdivide the concept of EROI into four distinct ideas: there is *mine-mouth* EROI ($EROI_{mm}$), *point-of-use* EROI ($EROI_{pou}$), *extended* EROI ($EROI_{ext}$), and *societal* EROI ($EROI_{soc}$).[4] The basic concept of EROI is the amount of energy gained from a source divided by the amount of energy required to access and use this source:

EROI = Energy contained in a source ÷ Energy required to get that energy

There are three subtypes of EROI because of an ambiguity in the notion of "get.: In particular, where does one "get" that energy "to"? The energy required to locate and dig coal to and get it as far as the mouth of the coal mine (or the head of the well, etc.) yields $EROI_{mm}$:

$EROI_{mm}$ = Energy contained in a source ÷ Energy required to get that source to the mine mouth (well head, etc.)

Getting the coal to the mine mouth, of course, is of little use in itself. It now needs to be transported—to a coal-fired power station, for example, and burned, which will engender additional energetic costs. This yields the idea of point-of-use EROI:

$EROI_{pou}$ = Energy returned to society ÷ Energy required to get and deliver that energy

However, transporting the coal to the power plant requires the construction and maintenance of roads, bridges, and motor vehicles, and these, too, will have energetic costs. The energy provided by burning the coal needs to be

---

[4] David Murphy and Charles Hall, "Year in Review—EROI or Energy Returned on (Energy) Invested," *Annals of the New York Academy of Sciences* 1185, Issue: *Ecological Economics Reviews* (2010): 102–118.

distributed to users, which will require the right sort of energy infrastructure, and so on. Adding these costs in yields *extended* EROI:

$$\text{EROI}_{ext} = \text{Energy returned to society} \div \text{Energy required to get, deliver, and use that energy}$$

Finally, there is also the concept of *societal EROI*—EROI_{soc}. This is the weighted, aggregate EROI of all energy sources employed by a society—that is, a function of the EROI of each source and the percentage of the society's energy needs that are met by this source. For example, if a society gets 60% of its EROI from coal, with an EROI of 30:1 and 40% from oil with an EROI of 20:1, then the societal EROI is $(30 \times 60\%) + (20 \times 40\%)$, yielding a societal EROI of 26:1. Being a different way of dividing up the conceptual space of EROI, societal EROI is less important for current purposes. The concept of societal EROI is, itself, subject to the distinction between EROI_{mm}, EROI_{pou}, and EROI_{ext}. That is, we can distinguish mine-mouth (well-head, etc.) societal EROI, point-of-use societal EROI, and extended societal EROI. These different versions of EROI are equally legitimate. But they are distinct, and when assessing the EROIs of different energy sources, it is crucial not to run them together. The result of this would be holding up different sources to different standards.

In a paper published in *Nature Energy* in July 2019, Paul Brockway and colleagues argued that this, in effect, is what routinely happens in the comparison of the EROIs of fossil fuels and those of renewable sources. Brockway's argument is based on a conceptual subdivision of the idea of EROI that is somewhat different from that of Murphy and Hall. Murphy and Hall's tripartite distinction between mine-mouth, point-of-use, and extended EROI has been replaced by a bipartite distinction employed by Brockway and colleagues. They argue that the standard estimates of the EROIs of fossil fuels have been artificially inflated by a consistent failure to distinguish between what they call *primary* and *final* energy stages.[5]

---

[5] P. Brockway, A. Owen, L. Brand-Correa, and L. Hardt, "Estimation of Global Final-Stage Energy-Return-on-Investment for Fossil Fuels with Comparison to Renewable Energy Sources," *Nature Energy* 4 (2019): 612–621. https://www-nature-com.access.library.miami.edu/articles/s41560-019-0425-z#Sec2. It would be an interesting conceptual exercise to work out exactly how Brockway's bipartite distinction maps on to Murphy and Hall's tripartite alternative. Mine-mouth EROI clearly corresponds to primary EROI, and final corresponds to extended EROI. The tricky part is working out which elements of point of use EROI get incorporated into the category of final EROI, and which into that of primary EROI. However, this sort of conceptual mapping is not required for my purposes.

Brockway's argument builds on work of Marco Raugei, published in the same journal 6 months earlier, that argues there is an endemic unfairness in the way EROIs of fossil fuels are calculated compared to those of renewable energy sources.[6] For example, as we shall see, one of the standard complaints against renewable sources such as wind or solar power is that they are *intermittent* sources of energy. The sun doesn't always shine and the wind doesn't always blow. If we want energy when our sources are being, in this way, uncooperative, we are going to find ways of storing it—in the form of batteries, for example—and this will add greatly to the energy cost (a lot of energy will go into the building, maintenance, and eventual decommissioning of these batteries), thus lowering the EROI of the renewable source. If we calculate the EROI of wind or solar in this way, then we are operating with what, in Brockway's language, is a *final* energy stage conception of EROI. The final energy stage is where the energy enters into the economy in immediately usable form. There is nothing wrong with employing a final energy stage assessment of EROI. It is an entirely reasonable way of assessing a source's EROI: what we are interested in is, precisely, how much of the energy of this source we can use. However, as Raugei points out—and Brockway and colleagues further develop this idea—calculations of the EROIs of fossil fuels are based not on a final, but on a *primary*, energy stage analysis.

Suppose you find a certain quantity of a fossil fuel, let's continue with coal, which will contain a certain quantity of energy. Of course, finding the coal and digging it up will have required a certain amount of energy on your part, as will transporting it to wherever you plan to extract the energy. The primary EROI will be the amount of energy contained in the coal divided by the amount of energy it takes to locate, transport, and burn the coal (including a percentage of the amount of energy that has gone into building the transport infrastructure, the plant, factoring in the lifespan, maintenance costs of both, etc.). However, burning coal does not, in itself, supply immediately usable energy into the economy. For that to happen, various other things need to happen. The energy created by the burning of the coal must be transformed and distributed. For example, the heat generated by the coal creates steam, the steam drives a turbine, and the turbine generates electricity. This electricity must then be transported to various portions of the grid before it enters the economy—that is, becomes immediately usable by consumers. There will be

---

[6] M. Raugei, "Net Energy Analysis Must Not Compare Apples and Oranges," *Nature Energy* 4 (2019): 86–88, https://www.nature.com/articles/s41560-019-0327-0.

energy embodied in these processes of transformation and distribution, and energy will have to be invested in building and maintaining the infrastructure required for this transformation and distribution. There will also be energy losses involved in this process of transformation. Only when we have subtracted these quantities of energy can the final EROI of the coal be calculated. The net energy of the coal is the energy contained in the coal minus all these other quantities of energy contained in the production, transformation, and distribution. And the final EROI is the net energy output divided by the total energy input, where the latter is equivalent to the energy in the coal *minus* all the energy involved in getting that energy to the consumer, where the resulting figure is then *divided* by all the energy involved in getting that energy to the consumer.[7]

The data for Brockway's calculations are supplied by two sources. The first is the extended world energy balances provided by the International Energy Agency for 142 countries and two rest-of-world regions. The second is the EXIOBASE MRIO database version 3.4, which provides input–output transaction figures that describe trade flows between 163 industries in 44 countries and 5 rest-of-world regions, plus details on final demand expenditure on each industry in each country/region.

The results of Brockway's calculations are quite startling. At the primary stage, the estimates more or less concur with the common wisdom that EROIs of fossil fuels lie around the 29:1 mark. However, at the final stage, the stage where the energy enters the economy and is available for consumption, the EROI, averaged out across all fossil fuels, is much lower—around 6:1. This figure, as we shall see, is comparable with many renewable forms of energy. Moreover, Brockway conducted a limited longitudinal analysis for the period 1995–2011. During this time, he argues, the primary EROI of coal dropped from around 50:1 to 29:1. Those of oil and gas dropped from the mid-to-high thirties to, as with coal, around 29:1. In the same time period, the final EROI of fossils fuels as a whole also went down, dropping from 6.8:1 in 1995 to 6.1:1 in 2011. Therefore, not only are the final EROIs of fossil fuels comparable to those of renewables, they are also declining.[8]

Brockway's calculations will, of course, be contested: the subject of counterclaim and counter-counterclaim, and so on. This is how it should be. In bruising contexts such as this, we would hope, the truth will eventually

[7] Brockway et al., "Estimation of Global Final-Stage Energy-Return-on-Investment," 615.
[8] Brockway et al., "Estimation of Global Final-Stage Energy-Return-on-Investment," 619.

win out. But, not having a crystal ball in hand, let's again use our method of moving from specific figures to the identification of some general trends, with the hope of unearthing some general principles. This, I think, is what we can, with relative safety, conclude.

First, the EROIs—both primary and final—of fossil fuels have been steadily and, a few occasional upticks in the case of coal aside, consistently, declining in recent decades. Second, this decline will continue as we continue to exhaust the easier sources of fossils fuels and must move on to their more-difficult-to-access counterparts. Thus, a further decline in EROI (both primary and final) is practically inevitable. Third, given their declining EROIs, there is already a legitimate question mark over the ability of fossil fuel sources to sustain a society, "recognizably similar to" ours. Fourth—and this really is the money ball—any sort of concerted, pervasive, or global use of CCS technology, of the sort required by the Paris Agreement, for example, will, in all likelihood, dramatically further lower the EROIs of fossil fuels. It seems it *must* do so, given the raw numbers involved. More carbon dioxide is produced by burning fossil fuels than the carbon that is contained in them. This carbon dioxide must be treated (liquefied by compression or cooling) in what are, energetically, very expensive ways. Storage and transport will also be energetically expensive. And eventual storage will also engender a series of costs and uncertainties. This is, perhaps, the most significant consideration because it is not factored into Brockway's calculations.[9]

We can conclude, I think without controversy, that the future of fossil fuels looks bleak. If we want to continue to use them, then we must try to capture their emitted GHGs, or climate calamity ensues. We don't even know if we will have the technology to do so on an industrial scale. We certainly don't have it yet.[10] But even if we did, this technology would dramatically decrease an EROI that is already declining due to the increasing difficulty of locating and accessing fossil fuel sources and may already be, at the final stage of analysis, no better than renewables. The EROIs of fossil fuels are going down anyway, and, for all practical purposes, may already be much lower than is generally recognized. Adding CCS costs to them will push them down even

---

[9] It couldn't be. Brockway is relying on figures from the International Energy Agency (IEA) and other bodies that were identified when CCS use was, at best, sporadic.

[10] Again, this is an issue on which claim and counterclaim abound. Proponents of renewables claim that CCS is useless. Proponents of CCS, however, claim it is the future. I am not getting involved. I think the prospects of CCS can be assessed independently of the issue of whether the technology actually does what it is supposed to. The central issue will be how much of a dent it puts in the already steeply declining EROIs of fossil fuels.

further. At some point, the threshold of viability will be breached. This is the dilemma of fossil fuels: the only way of making them viable, from a climate perspective, is by making them unable to sustain, from an EROI perspective, a society recognizably similar to ours.

It would, of course, be hubris for me, a philosopher, to adjudicate on the future of a developing technology such as CCS. Perhaps CCS will come through. Perhaps the difficulties adumbrated in these foregoing pages will be overcome. The conclusion that CCS cannot work would be premature, and I do not need this conclusion for the purposes of the argument I am in the process of developing. What I think we can legitimately conclude is that (1) given the already declining EROIs of fossil fuels, and (2) given the sorts of difficulties CCS faces at every stage of the process—capture, sequestration, storage, distribution, and deposition—it is *very far from clear* that CCS will do the job we need it to do. That is, it is very far from clear that CCS technology can reduce carbon emissions to the extent required while at the same time maintaining the EROIs of fossil fuels at the sort of level required to maintain society in a form recognizably similar to our own. That is to say, it would be unwise—foolish even—to assume CCS can do what we need it to do.

In its 2019, Special Report, the Intergovernmental Panel on Climate Change (IPCC) expressed medium confidence on the prospect of keeping global temperatures within the 1.5°C threshold ideal. That is, they expressed medium confidence that this threshold would not be breached by emissions alone.[11] However, this optimism is purchased by way of several assumptions—all of them rather radical. One of these includes widespread—indeed, global—afforestation programs, which will shall talk about later on in some depth. Another assumption, however, is that CCS technology will rapidly and dramatically improve by mid-century. As they put it: "Pathways limiting global warming to 1.5°C with no or limited overshoot would require rapid and far-reaching transitions in energy, land, urban and infrastructure (including transport and buildings), and industrial systems (high confidence). These systems transitions are unprecedented in terms of scale, but not necessarily in terms of speed, and imply deep emissions reductions in all sectors, a wide portfolio of mitigation options and a significant upscaling of

---

[11] The write: "but these emissions alone are unlikely to cause global warming of 1.5°C (medium confidence)," https://www.ipcc.ch/site/assets/uploads/sites/2/2019/06/SR15_Full_Report_High_Res.pdf A.2.

investments in those options (medium confidence)."[12] This is, as yet, more a hope than a justified belief. As Rob Jackson, an earth scientist at Stanford University, and Chair of the Global Carbon Project puts it, in an interview with *The Atlantic*: "Even some [of the scenarios] for 3 degrees Celsius assume that at some point in the next 50 years, we will have large-scale industrial activities to remove greenhouse gases from the atmosphere . . . It's a very dangerous game, I think. We're assuming that this thing we can't do today will somehow be possible and cheaper in the future. I believe in tech, but I don't believe in magic."[13]

A Nuclear Future?

Nuclear power is not a renewable technology. It relies on radioactive raw materials, such as uranium, that are not renewable. But they are plentiful. And, geopolitically speaking at least, they are also—unless, for example, Australia has an unexpected change of political heart—relatively easy to access. So, in this respect, they are far more like renewables than are the increasingly difficult-to-access fossil fuels. Moreover, like renewables, and unlike fossil fuels (who are universally accepted to have once had astonishingly high EROIs), the EROI of nuclear technology has always been a matter of dispute. Estimates of the EROI of nuclear power vary from less than 1 to more than 70. It might be no surprise to learn that the latter estimate is courtesy of the World Nuclear Association, an arm of the nuclear power industry.[14] That's quite a range. With an EROI of 1, nuclear power is entirely useless. With an EROI of 70 or more, it is, the odd Chernobyl-or-Fukushima-style incident aside, the savior of the planet. Even if we restrict ourselves to whole numbers, and if we assume that 1 to 70 is the full gamut of available estimates, this means that 69 of those possible estimates must be wrong. We need not overly speculate on the source of this falsity. The fact is that almost all of the claims concerning the EROI of nuclear energy, high or low, are false. How does one reason when almost all of the data on which one's reasoning and decisions will be based is not just erroneous but massively so?

---

[12] https://www.ipcc.ch/site/assets/uploads/sites/2/2019/05/SR15_SPM_version_report_LR.pdf, C.2.

[13] R. Meyer, "Are We Living through Climate Change's Worst-Case Scenario?" January 15, 2019.

[14] World Nuclear Association, https://www.world-nuclear.org/information-library/energy-and-the-environment/energy-analysis-of-power-systems.aspx.

The most obvious first step is to exclude the estimates that were not peer-reviewed—that is, were not published in peer-reviewed journals. This takes away the >70 and the <1 range of estimates. The problem with this approach, in this specific context, is that it does not leave us with a lot left in between. Murphy and Hall bemoan the lack of peer-reviewed work dealing with the calculations of EROIs for various energy sources.[15] This is especially true of nuclear energy. Perhaps the most systematic and comprehensive study to date is still that of Lenzen, who conducted a meta-analysis of the literature, averaging out various estimates of the EROI of nuclear fission. This yielded an EROI of 5:1.[16] If this figure is correct—and it may be worth putting an asterisk next to it because of the limited amount of data—then nuclear fission simply does not, as things stand, command a sufficiently high EROI to maintain a society recognizably similar to ours. A similar meta-analysis of 15 publications conducted by Lambert, Hall, and colleagues yields a somewhat more optimistic estimate of 14:1.[17] Weissbach et al.'s figure of 75 is very much an outlier.[18] Given the available evidence, it seems the most likely mid-range estimates for the EROI of nuclear power lie somewhere in the region of 5–14:1. This conclusion, however, must remain somewhat tentative.

A second step one can take in the face of this sort of disagreement is to follow our earlier practice of retreating from specific figures to general principles. The first general principle is that, on most peer-reviewed estimates, the EROI of nuclear power has always been relatively low compared to that of fossil fuels in their heyday. This is because of the truly massive start-up costs involved in building new nuclear plants, the massive reconditioning or upgrading costs and, to a lesser extent, the costs involved in the decommissioning of plants, the costs of uranium extraction and enrichment,

---

[15] Commenting on the publication of results of an intensive study of the literature they write: "Although each posting evoked massive amounts of commentary, essentially no new peer-reviewed papers were offered. There were many off-the-cuff calculations of much higher or lower EROIs for someone's favorite or least-favorite fuel (especially some very high numbers for nuclear), but there was little new peer-reviewed quality work." See "Year in Review—EROI or Energy Returned on (Energy) Invested," 110.

[16] Manfred Lenzen, "Life Cycle Energy and Greenhouse Gas Emissions of Nuclear Energy: A Review," *Energy Conversion and Management* 49, no. 8 (2008): 2178–2199. https://www.sciencedirect.com/science/article/pii/S0196890408000575?via%3Dihub

[17] J. Lambert, C. Hall, S. Balogh, A. Gupta, and M. Arnold, "Energy, EROI and Quality of Life," *Energy Policy* 64 (2014): 153–167. See also C. Hall, J. Lambert, and S. Balogh, "EROI of Different Fuels and Implications for Society," *Energy Policy* 64 (2014): 141–152.

[18] This is an outlier that almost certainly results from the adoption of a primary, rather than final, stage analysis of EROI. D. Weissbach, G. Ruprecht, A. Huke, K. Czerski, S. Gottlieb, and A. Hussein, "Energy Intensities, EROIs (Energy Returned on Invested), and Energy Payback Times of Electricity Generating Power Plants," *Energy* 52 (2013): 210–221.

and the costs of nuclear waste storage. These costs may vary somewhat over time. For example, the switch from diffusion to gas centrifuge may lower the costs of enrichment. However, in general, efforts to substantially reduce the costs of nuclear power have failed. For example, with regard to start-up costs, the Toshiba Westinghouse AP1000 was supposed to provide an easy-to-build, cheaper future for nuclear power. In fact, it turned out to be incredibly expensive, and Toshiba Westinghouse, with an annual loss from its nuclear projects estimated to be around $9 billion, was forced to file for bankruptcy, and then sold by Toshiba.[19] Moreover, in the construction of new nuclear plants this century, a clearly discernible pattern of delays and cost overruns is evident. Consider, for example, the European Pressurized Reactor (EPR) reactor built by French firm Areva in Olkiluoto, Finland, which cost more than three times the original estimate and ran more than 5 years overdue.[20] This escalation of costs and delays is even more evident in the case of a similar EPR reactor being built in Flamanville, France. Construction began in 2007. It was due to be up and running in 2012. It is still being built, with a date late in 2022 now cited as the earliest feasible start date. Originally supposed to cost €3.3 billion, costs have now exploded—no pun intended—to €10.9 billion. Part of the problem, but only part, was that some rather worrying weak spots—described by the regulators as "very serious"—were discovered in the reactor.[21] As a first general principle, therefore, I think it is reasonable to assume that, because of the immense financial costs involved—costs that ultimately translate into energy—whatever the specific figure for the EROI of nuclear fission turns out to be, it is likely to be considerably lower than that of fossil fuels in their golden days.

In the eyes of many, fusion, rather than fission, is the future of the nuclear industry. It may well be. That would be a great result for everyone. But, whether ultimately successful or not, fusion will be too late to address the climate issues we face now. Probably the best current hopes for fusion are the ITER Tokamak and the Spherical Tokamak. However, a nongenerating prototype of either of these is not expected to be in operation until 2035.[22] If the prototype works, then one might expect an early commercial version of

[19] "Westinghouse Files for Bankruptcy," *Nuclear Engineering International*, March 29, 2017.

[20] J. Rosendahl and T. Forsell, "Areva's Finland Reactor to Start in 2019 after Another Delay," *Reuters*, October 9, 2017.

[21] B. Felix and G. De Cleerq, "UPDATE 2-EDF's Flamanville Nuclear Plant Faces New Delay over Faulty Welding," *Reuters*, June 20, 2019.

[22] E. Gibney, "UK Hatches Plan to Build World's First Fusion Power Plant," *Nature*, October 11, 2019, https://www.nature.com/articles/d41586-019-03039-9.

this to be in operation by around 2050, perhaps later, by which time serious global warming would already be locked in.

There is one final, and I think reasonable, assumption to make. This is that there is little prospect of these nuclear start-up costs—certainly those of nuclear fission—being significantly reduced by technological innovation. While there might be some reductions, these are unlikely to be of a game-changing nature. This assumption is based on a second general principle, one that concerns how technology develops. Nuclear power—at least of the fission variety—is a 1970s technology. As such, we can expect it to already have reached maturity, or something close to maturity. In these circumstances, radical technological innovation should not be expected because this is not how technology works. As far as technology is concerned, middle age and innovation don't really go together. The reasons for this will also prove important for our assessment of renewables, and so we need to look at them in some depth.

The Diminishing Returns of Technology

On December 17, 1903, the Wright Flyer made the first controlled and sustained flight by a heavier than air powered aircraft. A few months shy of 66 years later, Neil Armstrong became the first man to walk on the moon. Sixty-six years, that's all, from the Wright Flyer to Apollo 11. In the 50 years since the moon landing there has been progress, certainly: in certain respects, substantial progress. But it has been nowhere near as dramatic as those first 66 years of powered flight. This is a common feature of technological development.[23] In technological development there is an initial breakthrough phase, where huge strides are made, the sorts of strides that would take you from the beaches of North Carolina to the surface of the moon in not much more than a half century. But, after this initial phase, progress begins to slow and continues to slow. Each new innovation, each new advancement, requires more and more work.[24] And eventually a point of diminishing returns is reached. The costs of further discovery outweigh the benefits. Investment ceases.

---

[23] The following discussion is indebted to J. Tainter, "Problem Solving: Complexity, History, Sustainability," *Population and Environment* 22, no. 3 (2000): 3–41.

[24] As the philosopher Nicholas Rescher (quoted in Tainter, "Problem-solving") put it: "In natural science we are involved in a technological arms race: with every 'victory over nature' the difficulty of achieving the breakthroughs which lie ahead is increased." See his *Unpopular Essays on Technological Progress* (Pittsburgh: University of Pittsburgh Press, 1980), 97.

Progress grinds to a halt. Sometimes, due to commercial factors, technology may even regress in certain respects.[25]

This characteristic trajectory of technological development is a corollary of the analysis of complexity and collapse defended by Tainter. Technological development is the result of a complex, communal process that evolves in a direction of increasing complexity. In the nineteenth and early twentieth century, science progressed through the efforts of lone wolf geniuses, such as Charles Darwin and Albert Einstein. Unfortunately, as Tainter points out, these geniuses put themselves out of business by answering all the questions that one can answer on one's own—as long as one is sufficiently brilliant, that is. What remains, after general questions such as the principles of natural selection, or of gravity, are solved are questions that are far more specialized and correspondingly more technically difficult to resolve. The result is that individual scientists must be replaced by teams of scientists: research teams.

The advent of research teams marks an increase in organization. Different scientists now have different roles within the team, just as different people came to take on different roles in our imagined coal mine in Chapter 4. These research teams will require costly equipment, whose care and operation requires technicians, whose purchase requires financial administrators, and whose management involves adherence to rules devised and put in place by government bodies, and so on and so forth. This increase in complexity is required to try to secure innovations that are becoming increasingly difficult to achieve.

Thus, fields of technological development all tend to follow a characteristic trajectory: from questions that are generalized to ones that are specialized, from lone wolf geniuses to larger and larger research teams, from simple to complex, and from low to high societal costs. Financially speaking, this is a costly trajectory. As a complex, communal enterprise, technological innovation is subject to the principle of diminishing returns. As a technology reaches maturity, the benefits of further innovation slowly become balanced, and eventually outweighed, by the costs.[26]

---

[25] Think, for example, of the fate of Concorde. Or, today, the Airbus A380.

[26] Good examples (provided by Tainter, "Problem Solving: Complexity, History, Sustainability") include the lineage of bombers produced for the US Air Force. In the 1950s and 1960s, 744 B-52 bombers were produced. In the 1980s, only 100 B-1 bombers—the replacement of the B-52, were made. In the 1990s, only 21 B-2 Stealth bombers—the replacement of the B-1 bomber were produced. This is the result of the increased complexity, and therefore cost, of the technological research program. This trajectory is a general feature of the trajectory of research development. Or consider, for example, medicine. The productivity index of medical research is defined as the average life expectancy of the population divided by national health expenditures considered as a percentage of gross national product. The graph that plots this relation year by year is almost like a cliff face—of

If this is correct, the implications for nuclear fission are clear. Commercial nuclear fission plants have been around since the 1950s. It is very likely that nuclear fission is already a mature technology. It would, therefore, be unrealistic to expect that it will be saved from its crippling start-up and other associated costs by further radical technological innovation. That is not to say it is impossible that this should happen. But given what we know about the general trajectory of technologies, it would be *irrational* to assume that it will. If this is correct, it is rational to assume that the crippling costs of nuclear power will likely always keep its EROI relatively low—quite possibly, too low to sustain a society "recognizably similar" to ours.

As we saw in the case of CCS technology, I choose to formulate claim in terms of the concept of doubt rather than assertion. A straightforward, if somewhat cautious and circumscribed, assertion would be: The preponderance of evidence suggests that the EROI of nuclear fission is too low to sustain a society recognizable similar to ours. My claim is slightly different, and even more cautious: As things stand, it is *very far from clear* that nuclear power will be able to supply an EROI capable of sustaining a society recognizably like ours. Perhaps it will, ultimately. But, as things stand, given the *seriousness* of the stakes involved, and given the current *evidential circumstances*, it would be unwise—indeed, irrational—to make this assumption. Serious doubt about the EROI of nuclear power is all I need for my dialectical purposes—that is, for the purpose of the argument I am developing in this chapter. And there is more than enough doubt about this to go around.

The EROIs of Renewables

The Empire of Fossil Fuels, at its height, marked a time of extraordinarily high EROIs. Henceforth, for ease of exposition, I shall express EROI values simply in whole numbers rather than in ratios. That is, in expressing EROI values, I shall assume that the denominator is 1, and simply list the numerator. In the 1940s, the EROI of oil was roughly 100. By the 1970s, it had declined to 23, but enjoyed a revival in the 1980s. Today it stands somewhere

---

the discouraging, downward variety. The same kind of decline is found in almost all research areas, including computer technology.

in the region of roughly 25–29.[27] This is a primary stage—well-head—figure. If Brockway and colleagues are correct, the final stage EROI is more in the vicinity of 6.1, having fallen from 6.8. These declining primary and final stage EROIs are largely because the easiest-to-obtain sources of oil were used up first, and we have now had to move on to wells that—either technically or geopolitically—are more difficult to access. As we saw earlier, it is inevitable that CCS restrictions, which we are going to have to place on oil-fired plants if we want to get climate change under control, will inevitably further reduce the EROI of this fuel substantially. The tar sands of Alberta are an often-touted alternative source of fossil fuel but, due to substantial problems of extraction, exacerbated by a brutal climate, the EROI of tar oil is generally thought to be only around 2 and, therefore, essentially useless.[28] How do renewable forms of energy stack up against these figures?

First, consider biofuels. The EROI of corn ethanol stands somewhere between 1 and 3, making it effectively a nonstarter.[29] Biodiesel is similar, sporting an EROI of roughly 3. Neither of these figures is surprising when you consider that only 30% of the solar energy striking the upper atmosphere reaches the ground and, of this 30%, only an astonishingly paltry 0.023% actually goes into plant photosynthesis.[30] If we need to sustain an EROI of around 11–14 (or, better, c. 25) to maintain a society recognizably similar to our own, then it is clear that biofuels are not even in the game. We shouldn't waste any more time on them. Indeed, as we shall see later, from a climate change perspective, biofuels are actually worse than useless: they take up land that can far more usefully be assigned another use.

The EROI of solar power is the subject of varying estimates, and caution seems an appropriate watchword. Some place it at 1—rendering it entirely useless—whereas other estimates range as high as 10, rendering it not far off the level required to sustain a society recognizably similar to ours.[31] Not

[27] The figure of 25:1 is a worldwide figure, obtained from a major survey article commissioned by the Royal Society and published in 2014. See David J. Murphy, "The Implications of the Declining Energy Return on the Investment of Oil Production," *Philosophical Transactions of the Royal Society A* 372 (January 2014). 29:1, 1–19 29 is the calculation of Brockway we encountered earlier.

[28] See Murphy and Hall, "Year in Review—EROI or Energy Returned on (Energy) Invested."

[29] See Hall, Lambert, and Balogh, "EROI of Different Fuels and the Implications for Society": "We believe that outside certain conditions in the tropics most ethanol EROI values are at or below the 3:1 minimum extended EROI value required for a fuel to be minimally useful to society."

[30] See J. Tainter, T. Allen, A. Little, and T. Hoekstra, "Resource Transitions and Energy Gain: Contexts of Organization," *Ecology and Society* 7, no. 3 (2003), https://www.ecologyandsociety.org/vol7/iss3/art4/.

[31] For the figure of 10, see J. Lambert, C. Hall, S. Balogh, A. Poisson, and A. Gupta, "EROI of Global Energy Resources: Preliminary Status and Trends," Report 1 of 2, UK-DFID 59717, November 2, 2012. P. Prieto and C. Hall estimate an EROI of 2–3 in *EROI of Spain's Solar Electricity System*

quite there, perhaps, but not far off. This figure was arrived by Hall, Lambert, and Balogh through calculation of the mean EROI value using data from 45 separate publications spanning several decades, and it may therefore be a promising one.[32] The same kind of divergence in estimates is also found vis-à-vis wind power, with EROI estimates ranging from 3 to 18.[33] The EROI of hydropower is also the subject of varying estimates, ranging from 10 to 84. The high value was calculated by Lambert, Hall, and Balogh, based on analysis of data from (what is admittedly) a relatively small sample class of 12 publications. Wave power is also estimated to have an EROI of around 15, which is quite promising.

The picture that emerges is a clear, but hardly definitive, one. Once we jettison biofuels, which we must, the principal remaining renewables are solar, wind, hydro, and wave. Then, the estimated EROIs of the four remaining renewable forms of energy are as follows:

Solar: 1–10

Wind: 3–18

Wave: 15

Hydro: 10–84

The upper-end estimations, taken at face value, make wind, wave, and hydropower viable forms of energy—if we assume sustaining a society recognizably like ours requires an EROI of 11–14—and place solar on the cusp of viability. The lower-end estimations render solar, wind, and probably even hydro incapable of sustaining our society in its current form. The first questions we have to address, therefore, are as follows: (1) Why is there is such variability in estimations? and (2) Which estimations—low-end or high-end—are more trustworthy?

To some extent—but by no means all—the immense variability of estimations derives from different initial assumptions concerning the scope

(New York: Springer, 2012). Similar estimates are to be found in D. Weissbach, G. Ruprecht, A. Huke, K. Czerski, S. Gottlieb, and A. Hussein, "Energy Intensities, EROIs (Energy Returned on Invested), and Energy Payback Times of Electricity Generating Power Plants," *Energy* 52 (2013) 210–221..

[32] Charles Hall, Jessica Lambert, and Stephen Balogh, "EROI of Different Fuels and the Implications for Society," *Energy Policy* 64 (2014): 141–152.
[33] The 18:1 estimate comes from a meta-analysis performed by C. Kubiszewski, P. Cleveland, and P. Endres, "Meta-analysis of Net Energy Return for Wind Power Systems," *Renewable Energy* 35 (2010): 218–225.

of energy investment, especially pertaining to whether we should be concerned with (in Brockway's terminology) primary or final stage EROI. To the extent we think of energy investment as the total amount of energy required to make energy immediately available to the consumer, we will be forced to move away from the higher-end estimations. The estimates of EROI values required to sustain a society recognizably similar to our own provided by Lambert, Hall, and colleagues are for primary stage EROIs.[34] Restricting ourselves to primary stage analysis, we might, not unreasonably, assume EROI values in the higher quadrant of the above ranges. But it is doubtful we could reasonably assume values right at the top of the range. Thus, for solar we might reasonably assume primary stage EROI of 5–10. For wind, we might assume 9–18. For wave, given the absence of a range, we have 15. For hydropower, somewhere in the 40–84 range might be justifiable.

For solar power, the final stage energy considerations that reduce EROI include its *intermittency*, which would require energetically (and financially) expensive storage solutions (in the form of batteries, for example). Another is *depreciation*, and the anticipated problems of ridding ourselves of vast quantities of degraded photovoltaic cells. For wind, final stage considerations—which, like solar, include intermittency and to a lesser extent depreciation—might compel us to move away from the higher-end estimates. But even if we restrict ourselves to the primary stage analysis, solar (5–10) and wind (9–18) appear uncomfortably close to the edge of the net energy cliff. Admittedly, wind looks in slightly better shape than solar, but even wind is, at best, on the cusp of viability. Remember, we are looking for EROIs of 11–14 if we want to sustain a society recognizably similar to our own, and ideally much higher if we want to preserve certain hallmarks of liberal democracy. There is nothing definitive about the estimated EROI ranges of solar or wind—nothing that positively screams out that they are going to make it.

Hydropower and wave power, ostensibly the two most promising forms of renewable energy, raise other issues. The problem with hydropower, as a general, climate-friendly solution to our energy needs is a straightforward one: in the developed world at least, the most suitable sites for hydropower stations have already been taken. Therefore, it is doubtful it can provide a *general* solution to our energy needs. We simply don't have enough places to

---

[34] Or, in the terminology of Murphy and Hall ("Year in Review—EROI or Energy Returned on (Energy) Invested"), "mine-mouth" or "well-head" EROI. See Chapter 4 for discussion.

put enough dams. The attempt to force hydroelectric solutions into progressively less and less suitable sites would inevitably result in declining EROI. Also, it will do a lot of environmental damage. You have to flood river valleys to create hydropower, destroying flora and fauna, sometimes rendering species extinct, and adversely changing the flow of rivers which result in further environmental impacts downstream. Even if it could provide all the energy we needed—which it almost certainly can't—the malign environmental impacts of hydropower would probably render it a not altogether desirable solution. If it can at all be avoided, we don't want to save the climate by destroying nature.

Wave power has related drawbacks. Some of these are environmental. If wave power were to be in serious contention as a general solution to our energy needs, we would have to be willing to substantially alter the coastline, and this is unlikely to be for the better. First, there are impacts associated with construction of the plant. Wave energy devices are typically tethered to the sea floor, by way of concrete pilings or blocks. Site preparation would require dredging and scouring of the sea bed to install electrical cables. The bigger the plant, of course, the greater the impact. There would also be impacts on marine creatures, especially mammals. There is some evidence of malign impacts on cetaceans and seals—certainly during the construction phase but also possibly during the operating phase. More research on this is needed.[35] Whales are also susceptible to entanglement in the mooring lines.[36] The plants might also act as barriers to marine movement and migration. Third, while wave power plants produce no GHG emissions, substantial GHG emissions are involved in their construction and transport, which would prove important should such plants became widespread. Fourth, there is the likelihood, at some point in the plant's life cycle of emission of toxic substances such as hydraulic fluids, lubricating fluids, and bi-fouling paints in the ocean. Fifth, wave power plants, especially on-shore or near-shore versions, can augment coastal erosion due to their alteration of currents and wave patterns. Slower or reduced ocean currents can also result in increased deposition of sediment. Sixth, they will be regarded by many as eyesores—largely because they are, indeed, very ugly.

---

[35] For a general overview, see M. Witt et al., "Assessing Wave Energy Effects on Biodiversity: The Wave Hub Experience," *Philosophical Transactions of the Royal Society A* 370 (2012), 502–529. https://doi.org/10.1098/rsta.2011.0265.

[36] See, for example, G. Boehlert and A. Gill, "Environmental and Ecological Effects of Ocean Renewable Energy Development," *Oceanography* 23 (2010): 68–81.

Most pertinently, given the type of argument developed in this chapter, compared to other renewables, wave power is in a very early stage of development. This makes an accurate assessment of costs difficult, since we have no definitive idea of the lifespan of the technology and the costs of connecting wave farms to existing power grids. Without this, accurate estimation of the EROI of wave power is difficult to ascertain. As the estimated EROI of wave power technology, the figure of 15 should, therefore, be treated with extreme caution. At 15, the EROI of wave power is still hovering around the cusp of viability. It may be much higher. But, of course, it may turn out to be much lower. At present, the general consensus is that while the costs of wave power are currently very high, these can be expected to drop significantly with further development. The evident immaturity of the technology is also cause for optimism. But, as in the case of solar and wind, there is certainly nothing definitive in what we know so far that screams out that wave power will be the solution to our problems.

Renewables and the Problem of Diffuseness

There is a general problem lurking behind renewables is the issue of *diffuseness*. In fossil fuels, solar energy is strongly concentrated. When readily accessible, the EROI of these substances is outstanding. The energy contained in renewables is much more diffuse.[37] In effect, fossil fuels and renewables are polar opposites. In the former, energy is highly concentrated but increasingly difficult to access. It is difficulty of access that is responsible for declining EROIs. In contrast, with renewables, the energy they provide is almost everywhere—sun, wind, and/or waves can be found in many places— but it exists in a highly diffuse or attenuated form. In effect, with coal, oil, and natural gas, nature has already done the job of concentrating solar energy. But since, in renewable energy sources, the energy is diffuse, we must do the work of concentrating it. To concentrate this energy, we must first collect it, and since the energy is diffuse, this means that our collection apparatus will need to be spread out over large areas. Thus, one consequence of the need to concentrate more diffuse forms of solar energy will be that we must turn over far more land to energy production. How much this will be will depend on

---

[37] This theme is developed by Tainter et al., "Resource Transitions and Energy Gain: Contexts of Organization."

what type of renewable energy source, or more likely combination of sources, we choose. But, in one way or another, and to some degree or other, our wide-ranging collection efforts will inevitably be environmentally destructive.

For example, the *MIT Energy Initiative* has calculated that powering 100% of the United States estimated electricity demand in 2050 with solar power would require roughly 33,000 square kilometers of land.[38] This is presented as a good thing. However, these calculations have been strongly contested, and they may have to be substantially revised upward.[39] But, suppose, for the sake of argument, they are right. In effect, an area somewhat larger than Massachusetts (27,000 square kilometers) would have to be turned over to energy production. This is only the area containing the solar farms them-selves. When we add in the surrounding infrastructure—access roads, bridges, extensions of the grid to the farms, and so on, the area involved will be substantially larger. This solution would, presumably, be problematic in smaller, densely populated, countries such as those of northern Europe. The United Kingdom, for example, has roughly one-sixth of the population of the United States but only one thirty-sixth of its land. Not to mention the fact that the sun doesn't shine very much there. But, for a large, and rela-tively non–densely populated country such as the United States, this might seem distinctly doable. It all depends on how you answer a particular ques-tion: How do you feel about losing an area of land larger than Massachusetts? Maybe twice the size of Massachusetts? Maybe even bigger than that?

For some, this would be a price worth paying. For others, it would be an environmental disaster. However, the question of size is of only indirect relevance. Size is an index of—another way of talking about—EROI. The lower the EROI of solar farms—a function of the diffuseness of the energy contained in solar radiation—the larger they will have to be. Conversely, higher EROI yields smaller farms. What matters directly is not the size of the farm but its EROI. We could cover the entire United States in solar farms. We could have solar panels on every roof in the country. The lower the EROI of solar, the more we would have to do this. But it still wouldn't matter if their EROI is not high enough. If it is not, they are not going to be able to sustain a society recognizably similar to our own. EROI is the North Star of this de-bate, and we should never lose sight of this.

[38] MIT Energy Initiative, https://energy.mit.edu/research/future-solar-energy/.
[39] MIT Technology Review, https://www.technologyreview.com/s/608126/in-sharp-rebuttal-scientists-squash-hopes-for-100-percent-renewables/.

Doubtifying and the Need for an Edge

Over the course of the last four chapters, I have slowly been building an argument. I haven't got to the punchline—more commonly known as the conclusion—of the argument yet. That is the subject of Chapters 6 and 7. Thus far, I have simply been putting in place the argument's premises. The first premise is:

P1. We need to mitigate climate change.

This was the subject of Chapters 2 and 3. We are not going to prevent climate change. There are enough GHGs in the atmosphere already to lock in substantial warming for some time. Without almost impossibly rapid action, it may already be too late for the Paris Agreement's stated goal of restricting warming to 2°C, but we still need to keep global temperature increase as low as we possibly can. The only way of doing this is by reducing the amount of atmospheric GHGs. Premise 1, therefore, commits us to at least this.

P2. We need energy sources with an EROI at least 11–14, and ideally much higher.

This premise was defended in Chapter 4. We do not have to accept Tainter's general account of societal collapse in order to accept this premise. Our society may or may not collapse, but even if it does not, the lowering of societal EROI is likely to result in society becoming noticeably grimmer: not just economically and in terms of quality of life, but possibly also politically.

The central concern of the current chapter has been to assemble Premise 3 of the argument. This premise concerns the likelihood of our available energy sources satisfying both of these conditions—low enough GHGs and high enough EROI. I prefer to put the premise in *doubtified* form. Yes, doubtified, from the verb to *doubtify*. It is a word. Or should be.

P3. There is substantial doubt as to whether any available energy source can satisfy the conditions specified in Premises 1 and 2.

To doubtify a claim is to take a straightforward assertion of fact and then transform it into a skeptical alternative. For example:

Assertion: The EROIs of fossil will drop below the threshold required to sustain a society recognizably like ours.

From this, we can construct its skeptical—that is, doubtified—cousin:

Doubtified: There is substantial doubt as to whether the EROIs of fossil fuels will remain above the threshold required to sustain a society recognizably like ours.

Or, switching from fossil fuels to a nuclear alternative:

Assertion: The EROI of nuclear fission is less than 15.

From this, we can construct its doubtified cousin:

Doubtified: There is substantial doubt as to whether the EROI of nuclear fission is as high as 15.

In other words, Premise 3 in the argument I am building is one constructed on uncertainty—upon what we don't know rather than what we do. The rationale for this lies, in part, in a desire to avoid disciplinary hubris. I am not a scientist. Who am I to say that the EROIs of fossil fuels will continue to decline below 11–14? Who am I to say that the EROI of nuclear fission, or of renewable forms of energy, is and/or is going to remain low? Perhaps someone is working on something right now that will solve these kinds of problems. I don't know.

Modern philosophy, however, was born in doubt—the methodological doubt of Descartes. I do know skepticism when I see it. I know doubt, and I am more than comfortable working with it. And I think the most reasonable interpretation of the morass of conflicting claims and counterclaims, figures and counterfigures, projection and denials of those projections, not to mention the paucity of research in certain critical areas, is a skeptical one. There is substantial doubt as to whether the EROIs of fossil fuels will remain viable, especially given the widespread use of CCS that is going to be required to satisfy Premise 1. There is substantial doubt over the EROI of nuclear fission. As things stand, we simply do not know whether its EROI is high enough to satisfy Premise 2.

There is similar substantial doubt over the EROIs of renewable energy sources. From the perspective of our current knowledge, I think the most reasonable conclusion is that things could really go either way. The potential triumph of renewables jostles with potential disaster, with the eventual outcome being too close to call. On one interpretation of the evidence, renewables seem to stand on the cusp of viability, and some may already have successfully ascended beyond this cusp. But on another interpretation, these renewables are far too close for comfort to the edge of the net energy cliff. Put in terms of the cliff imagery, we might think of biofuels as already having plunged into the abyss and now lie in a bloody mess on the canyon floor. Solar is teetering on the edge, with wind perhaps a step or two behind it. Hydropower and wave power are another step or two further back, but still far too close to the edge for those of nervous disposition. It might all turn out fine. But it could go the other way, too. And we just don't know enough—and can't see far enough into the future—to have much confidence. That, to me, seems the most reasonable interpretation of the current evidence.

I know doubt when I see it, and the literature on these issues is riddled with it, intentional or otherwise. I can doubtify Premise 2, if you like: there is substantial doubt as to whether a society recognizably similar to ours can be sustained with a societal EROI of less than 11–14. Hell, even Premise 1 can be doubtified. If we do not get climate change under control, then our future prospects, and those of the planet, are, to use a rather ominous medical term, *guarded*. That is, it is doubtful they are going to be unduly promising. Fossil fuels are, EROIcally speaking, clearly on the way down: there is substantial doubt over their future utility. Nuclear fission was always, using the same scale of measurement, a little questionable—that is, doubtful. But renewables—at least if we jettison those deadbeat biofuels—are not far off. Maybe they're nearly there. Maybe some of them are there—some of the time and in some places. Moreover, one thing that renewables do have in their favor is the relative immaturity of the technology—relative to fossil and nuclear technologies—which suggests the scope for improvement of EROIs is greatest in their case. If renewables really are on the cusp of viability, then we might profitably inquire if there is anything we might do to bolster them in the short term. If only there were something we could do to give our nascent renewables the edge. Is there such a thing? In the next chapter, I shall argue that there is. There is a step we can take to do this. Moreover, identifying this step is not exactly rocket science. But you're probably not going to like it. Not one bit.

6

The End of Meat

Collapse: A Brief Paean

You have worked your way through the family fortune and, quite frankly, had a lot of fun doing it. The weekends on the Cote D'Azur, the winter breaks in Tahiti. The private jet and the fleet of Ferraris. Who wouldn't envy you? I know I do, and I just made you up. But, alas, it's all gone. You've spent it all. Your life, as you know it, is coming to an end. No more snorting cocaine with superstars. No more mega-yacht in the Mediterranean or mansion in Monaco. Your life, you might be tempted to say—a little self-pityingly it has to be said—is *collapsing* around you.

Collapse: almost certainly one of the most difficult words in the English language to spin—along with *audit, tumor,* and *chlamydia*. He collapsed and died. She collapsed into a pit of despair. Their house collapsed into a sinkhole. It is difficult to make the word *collapse* sound appealing—unless you detest whatever it is that is collapsing, of course. The same is true of its cognates and synonyms: crumble, disintegrate, fold, flop, and so on. Collapse, it is tempting to think, is just an intrinsically bad thing for whomever or whatever is doing the collapsing. Except it is not. Not necessarily. Collapses can, in fact, be more or less graceful, and more or less prudently managed. Your fortune is gone, and you have choices to make. Choose well, and your life can still be comfortable, meaningful, and fulfilled. Choose badly, and you may end up sleeping in a cardboard box and panhandling at your local turnpike exit.

Like the Roman Empire, our empire of stored and concentrated solar energy is, in one way or another, coming to an end. There are two reasons for this. First, fossil sources of this energy are becoming ever more difficult to find and acquire. The resulting increased energy that must be put into finding and acquiring them is resulting in steadily declining energy returned on energy invested (EROI) values for these sources. Second, the byproducts of these fossil sources are so environmentally malignant that we can no longer afford to use them without effective forms of carbon capture and sequestration (CCS). Unfortunately, even if effective, industrial-scale forms of CCS

World on Fire. Mark Rowlands, Oxford University Press. © Oxford University Press 2021.
DOI: 10.1093/oso/9780197541890.003.0006

could be developed—and the jury is still very much out on that—they will be costly, financially and energetically, and, therefore, push the EROIs of these fossil sources further into problematic territory. This combination of factors will render fossil fuels nonviable solutions to our energy needs. Or, to put the matter in my preferred, doubtified (i.e., skeptical) form: there is substantial doubt over whether the EROIs of fossil fuels will remain in viable territory.

We have a reasonable idea of what a viable solution to our energy needs will look like. If the arguments of Lambert et al. are correct, we have reason for supposing that a societal EROI of below 25 is going to result in a moderate-to-poor quality of life. Once societal EROI drops below roughly 25, we will begin to lose certain hallmark features of successful liberal democracies: a healthy Human Development Index (HDI), healthy, well-nourished children, female literacy, gender equality, and the levels of health care expenditure to which we, in the affluent developed world, have become accustomed. Once societal EROI drops into the region of 14, then our ability to maintain features of society such as the arts (broadly construed) and even a well-functioning health system begins to dissipate. When we reach a societal EROI of 11, we find ourselves staring out into the abyss of the net energy cliff. Thus, 11–25 is the range we need to aim for in our societal EROI aspirations, and the closer we can get to 25 the better.

Until recently, fossil fuels effortlessly met, indeed far exceeded, this goal. But we can no longer rely on them. The evidence, sparse and contested though it is, casts considerable doubt on the ability of nuclear fission to get us to this goal. Most—not all, but most—of the peer-reviewed estimates seem to coalesce in the region of 5–14. It is to be hoped it is much higher than this, but there is substantial doubt over whether this hope will be fulfilled. Nuclear fusion, if it works, might certainly help us with our longer-term energy needs, but will almost certainly come too late to help us with our current climate woes. If this is correct, then it seems our remaining energy alternative would have to be of the renewable variety. Some of the proposed renewable sources are clearly not up to the job. Biofuels, in general—allowing for some variation with the category—sport unworkably low EROIs. This leaves us with solar, wind, hydropower, and wave power.

The problem with these latter energy sources is that they are entirely more *diffuse* than fossil or nuclear alternatives, or they have other features—such as the limited availability of hydropower sites—that make them unsuitable as a general solution to our energy needs. The evidence collectively paints a picture of these sources of energy. The picture is a little blurry in places, admittedly, but the general contours are, I think, clear. These renewable sources of energy stand on the *cusp* of viability. They may work—in the sense of supplying us with a sufficiently high level of societal EROI—but, equally, they may not. There is substantial doubt over whether they will do this. Things could go either way. The likelihood is that, in the short to middle term, we are going to have to learn to live with a significantly lower societal EROI than any of us, in the developed world at least, have ever really known.

When EROI dramatically declines, we can, if we so choose, flop around for a while, like a fish on a hook. This is, in effect, what the Romans did: devaluation, inflation, higher taxes, draconian social measures. But none of these things could stave off their collapse. Indeed, with the benefit of hindsight, always 20/20, we see that these were all just symptoms of this collapse. However, it doesn't have to go down as it did with the Romans. We have choices to make. In making these choices, I think it is important to bear in mind that, in the current state of the world, collapse may well be our best friend. As long as it is the right sort of collapse. There are far more elegant ways of collapsing than the thrashing-around-like-a fish-on-a-hook variety.

Collapse, in the sense employed by Tainter, simply refers to a *reduction in complexity*. Collapse is *simplification*. The complexity of a society is a matter of the number of distinct social roles it contains and the number of relations between them. The impressively complex society of the Romans—made up of soldiers, farmers, politicians, slaves, shoemakers, shipbuilders, gladiators, and many more—collapsed into something far simpler. The result was a few centuries of one of the oldest ways of life: subsistence farming. Prima facie, subsistence farming may not sound anywhere near as interesting as ancient Rome on a Saturday night—or *Nocte*, as I think they called it back then. And, to be quite honest, it isn't. But this doesn't mean that everyone is, necessarily, worse off after collapse compared to before. Just ask a gladiator.

Simplification is the name of the game we are going to have to play. But simplification comes in many forms: more or less graceful, more or less harmful. Indeed, some forms of simplification can be positively beneficial. In identifying the kind of simplification that would serve us best, it is always advisable to remember Heraclitus. All things are an exchange for fire, and fire for all things. The currency of the universe is energy, and it is the less-than-optimal energy choices we have made that have led us to the climate crisis. We are complex structures and, as such, need energy to survive. The same is true of the society in which we live. Old sources of energy—concentrated and stored solar energy in the form of fossil fuels—are no longer viable or soon won't be viable. We have new, renewable sources of energy, but it is far from clear that they are going to work—or work in the way and to the extent we need them to work. They will, we suspect, almost get us to where we want to be. But maybe not quite. Against this background, the general contours of our required course of action begin to emerge. The course of action we require is a precise targeting—and, ultimately, a simplification—of a certain section of the energy supply train that keeps us and our society in existence.

The Continuity of Food and Fuel, Again

As we saw in Chapter 5, Murphy and Hall distinguish between several forms of EROI: *societal* EROI, *extended* EROI, *point-of-use* EROI, and *mine-mouth* EROI.[1] The mine-mouth EROI of a quantity of coal is the energy contained in that coal as it leaves the mine divided by the amount of energy we had to put into getting it to this point. In the case of oil, we would, obviously, talk not of the mine mouth but the well head—yielding the idea of *well-head* EROI, the energy contained in a quantity of oil as it gushes out of the ground, similarly divided by the amount of energy we had to put into getting it to gush. But, in addition, Murphy and Hall also talk of *farm-gate* EROI—the amount of energy contained in an animal or crop as it leaves the farm on which it was raised or grown divided by the amount of energy that went into raising or growing it. Indeed, the notion of farm-gate EROI is, in their discussion, seamlessly conjoined with mine-mouth and well-head EROI. This is not a mistake but an assimilation, and one that is both appropriate and crucially important for my purposes. In discussing the EROI requirements of a society, it is important to understand an idea introduced in Chapter 1: the continuity of food and fuel. Both fuel and food are ways in which a society (and also individuals—but my focus is now on society) obtains the energy it needs to remain in existence.

At first glance, it may seem strange to lump chicken in with coal, and beef in with oil. But that's a consequence of thinking of the world as Heraclitus did. It is all energy in the end. The second law of thermodynamics entails that *any* complex structure—whether an organism or a society—needs to take in energy. If it doesn't, it will break down. We humans and our societies

---

[1]   D. Murphy and C. Hall, "Year in Review—EROI or Energy Returned on (Energy) Invested," *Annals of the New York Academy of Sciences* 1185, Issue: *Ecological Economics Reviews* (2010): 102–118.

are both complex structures and, as such, will break down in the absence of energy input. Some of this energy we need comes from fossil fuels, nuclear, and renewables—which keep humans alive by warming them, and by powering industry and technology that provides goods and services that, ultimately, keep humans alive and/or entertained. Some of the energy we need comes from food, which keeps us alive by allowing replacement of bodily cells and maintenance of the respiratory processes. In both cases, from an abstract vantage point at last, we are talking about the same *process* for the same *purpose*: an input of energy aimed at keeping a given complex structure in existence.

Therefore, from the Heraclitean perspective or, at least, on my particular spin on it, fuel and food are merely variants of the same thing: forms of energy that complex structures—such as humans and societies—use to remain in existence. There are, of course, superficial differences between food and fuel. Obviously, we ingest food but not fuel—the occasional gas-siphoning misadventure aside. Nevertheless, from the perspective of the first two laws of thermodynamics, what is crucial is that we absorb—take in—energy and not the specific method by which we do it. We take in the energy from food by eating it and breaking it down. But absorption of energy can take various forms and be achieved by different means. We absorb the energy from a fire, for example, through the transmission of molecular motion from it to us. Our fuel choices are continuous with our food choices, and vice versa. Both are, ultimately, energy choices. The common characterization of food as fuel is not inaccurate. Food is the fuel that drives the metabolic fire. Both food and fuel are part of our energy story—part of the story of what goes in versus what goes out—part of the story of *us*. Any point in the story—any segment of the chain of supply and consumption, whether it involves fuel or food—can be evaluated in terms of its EROI.

Stranger Things

Suppose there is an energy source of which you have become rather fond. In fact, it would be entirely accurate to describe you as a habitual user of this form of energy. Not just you, almost everyone—and for a long time, too. The only problem is that its EROI is nightmarishly upside down. You have to put far more energy into acquiring this source than the energy you ever ultimately get from it. That, surely, would make no sense. In the past, you could have compensated for this insanely inverted EROI component of your energy story by falling back on your empire of coal and oil, as a college student might fall back on the bank of mom and dad. But—oh no!—mum and dad are dead! A nasty business. Their superyacht sank off Porto Vecchio. Or was attacked by Somali pirates. Or something. But, whatever the truth of the matter, your debt is due and there is no one else to pay it. In these circumstances, your course is—surely—clear. You have to excise—cut out of your life—the part of your energy supply train that consists in the *upside-down* EROIs. What part of the chain, precisely, is that, you ask? Where, exactly, are these *Strange(r)* upside-down *Things* to be found? Okay. Okay. I've just lockdown binge-watched all three seasons of *Stranger Things* with my sons, and I am not going to talk about things being *upside-down* without getting in a reference.[2] However, it's possible that I digress. What, you might want to know—and this does get me back to my point—are these strange upside-downy EROI things? As I think I mentioned, in a little foreshadowing in which I indulged myself at the end of the previous chapter, the answer may be very unpopular.

Meat, that's what. *Meat* is the ultimate EROI inversion, with EROIs ranging anywhere from 1:4 to more than 1:30, depending on the meat in question and the assumptions made in the calculation. Meat is the clearest part of the energy continuum that needs changing. Meat is what we have to give up if we want our lives to go on in recognizably the same way, without losing too much of that arts-containing, health-care-providing, open and democratic society we have come to know and love.

In Chapter 7, I provided an analysis of various putative energy sources in terms of their EROI. In the rest of this chapter and the next, I am going to erect an argument that builds on that analysis. It is an argument for why we must give up our habit of eating meat, and it looks like this.

(1) There is no alternative to renewables. However:
(2) Renewables are, as judged by their EROI, on the cusp of viability at best. To give them the edge they need, we will have to do two things:
(3) Stop eating meat: stop dealing in stranger things—meat is the arch offender—that have upside-down EROIs. These place a now

---

[2] For those of you who have not seen *Stranger Things*, the *Upside Down* is a strange word, a dark, shadowy reflection of our own, where . . . Oh, forget it. Where have you been?

intolerable strain on the energy chain that keeps us in existence. And, if we stop doing this, then we can:

(4) Reforest the land now used for grazing or growing food for the animals we eat.

The overall result of steps (3) and (4) will be a healthier, more stable energy chain with far fewer symptoms of ill health—such as off-the-hook greenhouse gas (GHG) emissions.

I argued for Premise (1) in Chapter 5. I argued that the EROI of fossil fuels will, in all likelihood, decline to a point of nonviability—maybe not today, maybe not tomorrow, but soon—due to the twin pressures supplied by harder-to-access fossil sources combined with the necessity of using CCS technology, assuming it can be made to work. Nuclear fission, with an EROI of probably somewhere in the region of 5–14, is not a viable replacement—or there is, at the very least, substantial doubt over its viability—and given the maturity of the nuclear technology, there is no reason to expect any dramatic innovation that will push this EROI value up much further. Therefore, we arrive at (1). There is no alternative to renewables.[3]

In Chapter 5, I also argued for (2)—the claim that renewables are, at best, on the cusp of viability. The strikingly low EROIs of corn-based biofuels render them nonviable. They are not even in the game. Prospects are better for hydropower and wave power, while prospects are more guarded for solar and wind. If hydropower and wave power are a step or two back from the edge of the net energy cliff, then solar and wind are teetering on the brink, and biofuels plunged into the abyss some time ago and are now just an ugly mess on the floor of the canyon. As things stand, it simply is not clear that renewables are going to be up to the task of sustaining a society "recognizably similar" to our own. To make this more likely, we need to find a way of giving them an *edge*.

---

[3] Or, in my preferred skeptical form: there is substantial doubt over the abilities of fossil and nuclear power to meet our energy needs given the constraints posed by the need to arrest climate change. That is, there is (1) substantial doubt over whether the EROI of nuclear fission is high enough and (2) substantial doubt over whether the EROIs of fossil fuels will remain high enough given increasing scarcity and the necessity of using CCS technology. Admittedly, there is also substantial doubt about renewables, too. But, I have argued, probably the most consistent interpretation of the evidence places them on the cusp of viability—which is better than declining beyond that point or probably never having been at that point, especially given the relative novelty of renewable technology coupled with the way technology tends to develop. As I have mentioned, I prefer the skeptical—doubtified—version. But for ease of exposition, I shall develop this argument in straightforward assertoric form. Please feel free to render the argument in skeptical terms if you prefer.

Interestingly, at least for those who like arguments and their structure, from the perspective of the argument's validity, Premises (1) and (2) are not essential in their current form. Suppose, for example, you are unconvinced by renewables and think the future is nuclear. As long as you agree that the EROI of nuclear fission is, currently, only around the cusp of viability, then you are at liberty to modify the first two premises: (1*) There is no alternative to nuclear, and (2*) Nuclear, as judged by its EROI, is only on the cusp of viability. You could even, if you are so inclined, run a suitably modified form of argument with fossil fuel versions of (1) and (2). For example: (1**) There is no alternative to fossil fuels, and (2**) Due to declining availability and the necessity of using CCS technology, the EROIs of fossil fuels will soon diminish to the cusp of viability. The argument I am presenting is based on substantial doubt about our available energy sources to sport or maintain high enough EROIs to meet our energy requirements. The validity of the argument does not, in fact, require taking a stand on which type of energy source is the most promising. Why then, one might wonder, have I spent so much time arguing for renewables—that there is no alternative to them and that they, specifically, are on the cusp of viability? The reason is that, to work, an argument needs to be *sound* as well as valid—and for this the argument needs premises that are true. I think the most reasonable interpretation of the available evidence makes Premises (1) and (2) more likely to be true than (1*) and (2*) or (1**) and (2**). Nevertheless, as we all know from elementary logic class, a disjunction is never less true than any of its disjuncts. So, one could always disjunctivize the first two premises: *Either* (1) and (2) *or* (1*) and (2*) *or* (1**) and (3**). Indeed, disjunctivized in this way, I suspect the argument will be close to—not quite there, but close to—unassailable. Close to unassailable, perhaps, but unwieldy and bloody difficult to follow. Therefore, I shall continue to develop the argument in terms of my preferred (1) and (2). I simply wanted everyone aware of the logical possibilities should their preferred energy sources lie elsewhere. If you are unhappy with (1) and (2), then please feel free to disjunctivize away.

Moving on. In this chapter and the next, I shall argue that the edge we need to supply to renewables (or . . . insert one's under-pressure *source d'énergie de choix*) is a precise targeting of (a portion of) the energy supply chain that keeps us, and our society, in existence. This targeting consists in abandoning the animal agriculture industry and assigning another purpose to the land currently used for grazing and growing crops to feed livestock. It is, I shall argue, fulfilling the steps described in (3) and (4) that will push renewables far enough away from the edge of the net energy cliff and, so, make them a viable solution to the climate problem. It pushes them away from the edge of the energy cliff for the simple reason that *a significant part of the drain on renewable forms of energy will be provided by the requirement that they subsidize this insanely inverted part of the energy supply train that we call the animal agriculture industry.* Once these shackles have been released, these renewable energy sources can devote more of themselves to other parts of the energy supply train that we need to keep ourselves and our societies in existence.

In (3) and (4), therefore, we find the key to the problem of climate change and also—these are topics we shall address later—to the problems of pestilence and mass extinction. If the argument I shall develop is sound, these are not difficult problems to solve, in the sense that their solutions are not insuperable intellectual riddles and we won't need radical new technologies to solve them. The solutions to these problems are easy to identify but difficult to implement. They are problems of the will rather than the intellect. In this chapter I am going to defend (3). In the next chapter, I shall argue for (4).

The Consequences of Complication

The livestock industry. Animal agriculture. Two different names for what I am going to call *the meat industry*. It's an ugly label. But it is an ugly industry. The label isn't quite right, because it is intended to cover the production of dairy and other animal products, too. But it's near enough, and everyone knows what I'm talking about. The *meat industry* is the industry that raises animals so we may eat either them or their products.

It is important to realize that the meat industry is an unnecessary wrinkle in the energy supply train that keeps us and our society in existence and that doing away with it would, accordingly, be a *simplification*. We do not need to obtain our energy from animals. In fact, nothing that we acquire from animals or their products is necessary for our health. The American Dietetic Association, the premier organization for nutrition professionals in the United States, writes: "It is the position of the American Dietetic Association that appropriately planned vegetarian diets, *including total vegetarian or vegan diets*, are healthful, nutritionally adequate, and may provide health benefits in the prevention and treatment of certain diseases."[4] The British

---

[4] W. Craig and A. Mangels, "Position of the American Dietetic Association: Vegetarian Diets," *Journal of the America Dietetic Association* 109 (2009): 1266–1282, at 1266. Emphasis is mine. The

Dietetic Association agrees: "Plant-based diets are becoming more popular and if they are well planned, can support healthy living at every age and life-stage."[5] It is now widely accepted—indeed, a commonplace—that we can live healthy lives on a purely plant-based diet. A little bit of care and education is required, but this is certainly possible, and when done well, it can be a healthier alternative to a meat-based diet. The eating of meat—the part of our energy supply train that involves animals—is, therefore, a choice rather than a necessity. Taking the meat out of the equation—ingesting the energy from plants directly and cutting out the meaty middleman—is, therefore, a simplification. It is a *reduction in complexity*. This, I shall argue, is, *perhaps*, the only form of *collapse* that we need undergo in trying to solve our climate woes. And even if it is not the only form of collapse we might have to accept, it is by far the most significant one.

Choosing to eat meat introduces the kind of grossly inverted EROI values outlined earlier. We shall look at these figures more closely later. However, roughly speaking, we know that we have to put in anywhere between 4 and more than 30 times as much energy into producing meat as we acquire through eating it. For every kilojoule of energy that we acquire from meat, we will have invested at least 4, and maybe more than 30, kilojoules of energy. Since we don't have to eat meat, doing so is an unnecessary complication in our energy supply. The consequences of this complication are far-reaching and unfortunate.

It is well known that the meat industry produces a *lot* of GHG emissions: a significant portion of the overall amount of anthropogenic GHGs derives from this industry. According to the Food and Agriculture Organization (FAO) of the United Nations, the meat industry is responsible for 14.5% of all anthropogenic GHG emissions.[6] An earlier study by the FAO estimated the meat industry to be responsible for 18% of anthropogenic GHG emissions, but the FAO regards its newer estimate as more accurate.[7] Whether

---

Academy of Nutrition and Dietetics concurs, V. Melina and W. Craig, "The Position of the Academy of Nutrition and Dietetics: Vegetarian Diets," *Journal of the Academy of Nutrition and Dietetics* 116 (2016): 1970–1980, at 1970.

[5] The Association of UK Dietitians, https://www.bda.uk.com/resource/plant-based-diet.html.

[6] Food and Agriculture Organization of the United Nations, *Tackling Climate Change through Livestock: A Global Assessment of Emissions and Mitigation Opportunities* (2013), http://www.fao.org/3/i3437e/i3437e.pdf.

[7] Although it also acknowledges that, given the different reference periods and sources, it is difficult to make a like-for-like comparison. For the 2006 report, see Food and Agriculture Organization of the United Nations, *Livestock's Long Shadow: Environmental Issues and Options* (2006), http://www.fao.org/3/a-a0701e.pdf.

14.5% or 18%, that is more than the entire transport sector—planes, trains, automobiles, and ships—combined (which amounts to just over 13%). The newer, 14.5%, estimate translates into 7.1 gigatons of CO_2 *equivalent* per year.

A *gigaton*: that's a nice unit. It is equal to 1 billion metric tons of CO_2 equivalent per year. The notion of CO_2 equivalence is important since, as we shall see later, it has a bearing on how we count GHG emissions. The reason we need the idea of CO_2 equivalence is because carbon dioxide is not the only greenhouse gas. Carbon dioxide is the most notorious, but two other gases produced by the meat industry are also important: methane (CH_4) and nitrous oxide (N_2O). Carbon dioxide is the most infamous of the GHGs not because of its efficiency in trapping heat, but simply because there is much more of it around. According to the Environmental Protection Agency (EPA), 82% of anthropogenic GHG emissions consist in carbon dioxide, 10% consist in methane, and 6% in nitrous oxide.[8] However, as we saw in Chapter 3, although not as prevalent as carbon dioxide, methane and nitrous oxide are much more efficient GHGs—much better at trapping heat—than carbon dioxide. Carbon dioxide, by stipulation, has a global warming potential (GWP) of 1. On this scale, methane has a GWP of roughly 25. Nitrous oxide has an astonishing GWP of roughly 300. The notion of CO_2 equivalence is needed to give us a single scale of measurement for GHGs and their warming effects. Thus, if you release a ton of methane into the air, that has the same heat-trapping impact as if you released 25 tons of carbon dioxide. And if you released a ton of nitrous oxide into the air, that would have the same impact as if you had released 300 tons of carbon dioxide.[9]

Looking at the meat industry as a whole, feed production and processing, including change in land use, are responsible for 45% of total emissions. Enteric fermentation from ruminants is responsible for a further 39% of total livestock emissions. Yes, ruminants are extraordinarily gassy creatures. About as gassy as any normal human would be after six pints of Stella Artois and a Vindaloo. Unlike us, however, ruminants can't help it. Thus, cattle— whether raised for beef or dairy—are the type of livestock responsible for most emissions, representing over 60% of total livestock emissions. Manure storage and processing represent roughly 10% of the sector's emissions. The

---

[8] US Environmental Protection Agency, https://www.epa.gov/ghgemissions/overview-greenhouse-gases.
[9] US Environmental Protection Agency, https://www.epa.gov/climateleadership/atmospheric-lifetime-and-global-warming-potential-defined.

remaining emissions result from the processing and transport of animal products.[10]

If we switch from quantity of emissions to types of gas emitted, we find that roughly 44% of livestock emissions take the form of methane. The remaining gases consist in almost equal portions of nitrous oxide (29%) and carbon dioxide (27%). Viewed as a percentage of overall anthropogenic GHG emissions, this translates into the meat industry bearing responsibility for 5% of total anthropogenic carbon dioxide emissions, 44% of anthropogenic methane emissions, and 53% of anthropogenic nitrous oxide emissions.[11]

Emissions intensity figures prove very interesting, if you like that sort of thing. An emission intensity is the amount of CO_2-equivalent gas produced per unit of product. Emission intensity is highest for beef: almost 300 kilograms of CO_2-equivalent gases are released for every kilogram of beef protein. Next comes meat and milk from small ruminants (chiefly, sheep, and goats), with the meat of these creatures producing 165 kilograms of CO_2-equivalent gas for every kilogram of meat protein, and their milk releasing 112 kilograms for every kilogram of dairy protein. Cow milk, chicken products, and pork have slightly lower global average emission intensities, in the region of 100 CO_2-equivalent GHGs per kilogram.[12]

Let's play with some math, to put a little meat on the bones—the bad pun is, alas, entirely intended—of these figures. A portion of lean beef is generally regarded as having around 31 grams of protein per 100 grams of beef—so a little under a third of the total weight of the portion is made up of protein.[13] A kilogram of beef will, therefore, contain around 310 grams of protein. Half a kilo is around 150 grams. The 16 ounce New York Strip sitting on your plate is just under half a kilo—roughly 0.45 kilos or 450 grams. This will, therefore, contain roughly 140 grams of protein. A kilo of beef protein produces 300 kilograms of CO_2-equivalent GHGs. This means that for this steak to be sitting on the plate in front of you required the emission of 42 kilograms of CO_2-equivalent GHGs. On any given visit to their favorite steak house, a

[10] Food and Agriculture Organization of the United Nations, *Tackling Climate Change through Livestock: A Global Assessment of Emissions and Mitigation Opportunities*.

[11] Food and Agriculture Organization of the United States, *Tackling Climate Change through Livestock: A Global Assessment of Emissions and Mitigation Opportunities* (2013). For a summary, see http://www.fao.org/news/story/en/item/197623/icode/.

[12] Food and Agriculture Organization of the United States, *Tackling Climate Change through Livestock: A Global Assessment of Emissions and Mitigation Opportunities* (2013). For summary, see http://www.fao.org/news/story/en/item/197623/icode/.

[13] This is according to nutrition.org/uk, https://www.nutrition.org.uk/nutritionscience/nutrients-food-and-ingredients/protein.html?start=4.

family of six, all fond of especially large steaks, might easily be responsible for a quarter of a metric ton of GHG emissions.

Outliers Again

These emissions figures do seem somewhat disturbing. There are, however, higher estimates out there. One paper argues that the meat industry is responsible for 51% of emissions.[14] As you will know by now, as a general methodological rule, I think it is a good idea to view outliers with extreme suspicion, especially if they occur in non-peer-reviewed work. That being so, I think the best method is to begin with the FAO estimate. I'll use this as a baseline assumption. As a matter of further investigation, it might be useful to examine how the FAO arrived at this figure, with a view to seeing if there are any steps in their calculation that might be questionable. I think there may be one place in particular where the FAO's reasoning can at least be questioned. It pertains to how we calculate the CO_2 equivalence and resulting GWP of methane.[15]

Methane is a much more ephemeral gas than either carbon dioxide or nitrous oxide. It doesn't hang around in the atmosphere anywhere near as much as carbon dioxide. The *half-life* of a gas in the atmosphere is the amount of time it would take for half of a given quantity of that gas to disappear. If, for example, you release a ton of carbon dioxide into the atmosphere, then the half-life of carbon dioxide is the time it would take for half of that ton to disappear. Disappearance can take two main forms, degradation of the gas or its being absorbed in some other medium.[16] The half-life of carbon dioxide is a generally estimated to be around 100–200 years. Actually, it's a lot more complex than that as carbon dioxide has a very *long tail*. While 50% of accumulated carbon dioxide might degrade within 50 years, and 70% within

---

[14] See R. Goodland and J. Anhang, "Livestock and Climate Change: What If the Key Actors in Climate Change Were Pigs, Chickens and Cows?" http://www.worldwatch.org/files/pdf/Livestock%20and%20Climate%20Change.pdf. The paper became well known, in part, because of its being linked to on *Cowspiracy website*. See http://www.cowspiracy.com/.

[15] Goodland and Anhang—in the paper responsible for the 51% estimate—also question the FAO figure for precisely this reason. See R. Goodland and J. Anhang, "Livestock and Climate Change."

[16] The so-called *airborne fraction* of a GHG is the fraction of the gas that stays in the atmosphere rather than being absorbed in some other medium. The disappearance of at least some carbon dioxide is, in fact, no cause for celebration, for this simply means it has been absorbed into the world's oceans, rendering them more acidic. We may well have a desertification of the oceans time bomb to look forward to in the coming century. On this, see for example, Stephanie Renfrow, "An Ocean Full of Deserts," https://earthdata.nasa.gov/learn/sensing-our-planet/an-ocean-full-of-deserts.

a hundred years, the 200 years after will see only another 10% decline. And it can take 30,000–35,000 years to be rid of all of it.[17] We might be tempted to say that the half-life of atmospheric carbon dioxide is 50 years. And while this might, strictly be correct, it is also misleading since it diverts attention away from the fact that a not insignificant portion of the carbon dioxide we pump into the atmosphere can stay there for thousands of years. The half-life of nitrous oxide is generally regarded as being around 100 years plus.[18] But the half-life of methane is much shorter—only around 8.6 years.[19] This can affect how we calculate its GWP.

The crucial thing to realize is that CO_2 equivalence is a quantity that is assessed *over* time not *at* a time. That is, if we say that methane has a GWP of 25, this means that a given unit of methane will trap 25 times as much heat as carbon dioxide *over a given time period*. The question is: how do we choose this time period? The FAO assessment of the impact of the meat industry chose a 100-year time period. That is, their assumption was that, over a period of 100 years, a given unit of methane will trap 25 times as much heat as a given unit of carbon dioxide. The assumption of a 100-year frame is common practice, and so the FAO's assumption was not arbitrary. However, that it is common does not mean that it is useful in this case. The typical half-life of methane is 8.6 years. Therefore, if we assume a 100-year time frame for assessment, then we are assessing the heat-trapping ability of a gas that *won't be around for much of the time frame in which it is assessed*. That does seem a little odd. Suppose we change the time frame of assessment to one that more adequately reflects the time that a given quantity of methane will actually be in the atmosphere—say 20 years. If we do this, then methane will have a much higher GWP—somewhere in the vicinity of 72.[20] That is, over a period of 20 years, a given unit of methane will trap somewhere in the region of 72 times as much heat as the same unit of carbon dioxide. Therefore, if we assume a 20-year time frame for assessment, the CO_2-equivalence figures will jump quite dramatically.

[17] D. Archer, M. Eby, V. Brovkin, A. Ridgwell, L., Cao, U. Mikolajewicz, K. Caldeira, K. Matsumoto, G. Munhoven, A. Montenegro, and K. Tokos, "Atmospheric Lifetime of Fossil Fuel Carbon Dioxide," *Annual Review of Earth and Planetary Sciences* 37 (2009): 117–134.

[18] University of Warwick, "Elusive Compounds of Greenhouse Gas Isolated," *ScienceDaily*, September 17, 2019, www.sciencedaily.com/releases/2019/09/190917115439.htm.

[19] R. Muller and E. Muller, "Fugitive Methane and the Role of Atmospheric Half-life," *Geoninformatics and Geostatistics* 5 (2017), 10.4172/2327-4581.1000162.

[20] See S. Vaidyanathan, "How Bad of a Greenhouse Gas Is Methane? The Global Warming Potential of the Gaseous Fossil Fuel May Be Consistently Underestimated," *Scientific American*, December 22, 2015.

Which time frame is most justifiable to use in assessing CO_2 equivalence? Twenty years or one hundred? Or somewhere in between? Much will depend on how urgent you think the problems of climate change are. It used to be common to think that most of the serious problems of climate change would only emerge in the next century. If this were true, assuming a 100-year time frame might be reasonable. But, as we have seen, this assumption is highly questionable. The United Nations, for example, has placed 2030 as the year by which we need to reduce GHG emissions by 45% if we are to stay within the 1.5°C threshold.[21] The worrying possibility of tipping cascades, encountered in Chapter 3, might add further urgency to this assessment.

If we did adopt the 20-year time frame for assessing the GWP of methane, then the figure of 7.1 gigatons of CO_2-equivalent gases would have to be revised upward. This is inevitable, since a GWP of 72 is nearly three times higher than 25, thus raising the CO_2-equivalence figures. This, however, need not be reflected much, if at all, in the estimated percentage figures for the GHGs produced by the meat industry—since if we assume a 20-year time frame for calculating the GWP of methane emissions from the meat industry, then we would have to do assume the same time frame when assessing methane contributions from outside the meat industry (from coal, natural gas, etc.). Nevertheless, this stagnation of the percentage value should not blind us to the fact 7.1 gigatons per year may be somewhat conservative estimate of the CO_2-equivalent GHG emissions deriving from the meat industry. Or it may not. For my purposes, I can afford to treat this issue with equanimity. The figures of 7.1 gigatons and 14.5% of total anthropogenic GHG emissions will work perfectly well for what I plan to do with them.

Feed Conversion Ratios: Don't Deal with Mr. Beef

On the face of it, 14.5% of all anthropogenic GHGs attributable to the meat industry does seem a somewhat significant portion. And 7.1 gigatons of CO_2-equivalent gases doesn't seem to be the sort of thing that we can afford to, if you'll forgive the pun, sniff at. However, from the perspective of trying to fix them, or at least mitigate them, the most important first step is to see these figures for what they really are: *symptoms*. They are symptoms

---

[21] General Assembly, 73rd Session, High Level Meeting on Climate and Sustainable Development, March 28, 2019, https://www.un.org/press/en/2019/ga12131.doc.htm.

of a bad energy exchange: an energy investment that has gone awry. We are putting 7.1 gigatons of CO_2-equivalent GHGs—14.5% of all anthropogenic GHGs—into the atmosphere every year because we insist on converting energy contained in plants to energy contained in animals, and then harvesting the latter. This, from the perspective of EROI, is a hopelessly upside-down way of acquiring energy.

The upside-down character of the energy exchange is the result of the conversion ratio of plant matter into animal matter. Animals are notoriously inefficient food—that is, energy—converters. This is well documented and cannot really be disputed. However, precisely how inefficient is a matter of dispute. Yes—and, by now, I'm sure this won't come entirely as a surprise—estimates of what is known as *feed conversion ratio* (FCR) vary. A lot. Disingenuity, it seems, is common on both sides, and it is difficult to find nonpartisan estimates. Even if we confine ourselves to middle-range estimates—eliminating low- and high-end outliers, the estimates vary dramatically. For chickens, the estimates generally vary between 2× and 5×. That is, chickens eat somewhere between twice as much and five times as much food as they produce. For pigs, the figure is between 4x and 19x. For cattle, the figure rises to between 6× and 30×. Why is there this variation, even in the middle range of estimated conversion efficiencies?

The variation, to a considerable extent, stems from differences in what gets measured. A *live weight* feed conversion ratio takes as the basis of measure the entire weight of the animal before slaughter. Live weight ratio FCRs are generally used by the meat industry in producing their low-end figures. These figures, however, are misleading. There are various parts of the animal that we cannot eat because we are physically incapable of doing so (e.g., the fur and large parts of the animal's skeleton), and various parts that we don't eat because they would be disgusting (a chicken's ass sandwich, anyone?) or potentially harmful (cow brains—remember mad cow disease?). Therefore, it is more accurate—not to mention, honest—to employ in calculations what is known as an *edible weight* feed conversion ratio, which takes as the basis of calculations only the weight of the parts of the animal we actually eat.

Even with respect to the edible weight, however, there is a further ambiguity between carcass hanging weight and final take-home weight. Hanging weight is greater because certain things, such as water, will be lost in cooking. The final take-home weight is generally thought to be between one-half and one-third of the live weight (one-half for pigs, and one-third for cows). Using final take-home weight as the basis of calculation tends to push up the feed conversion ratio estimates toward the higher end of the middle-range spectrum, if you know what I mean. Thus, while the beef industry commonly cites a figure of a 6:1 conversion ratio for beef—that is 1 kilogram of beef for every 6 kilograms of feed consumed—estimates from those outside the industry, or opponents of the industry, tend to cite figures between 16:1 and 54:1. Much of this difference—not all, but much—derives from the decision of whether to use live weight or take-home weight as the basis of comparison. Since take-home weight does seem to be a more accurate measure of how much we are actually getting out of the animals we eat, it is reasonable to suppose that the middle-to-higher-end estimates are more reliable.[22]

Shepon and colleagues calculated both caloric and protein conversion efficiencies for livestock. When averaged out over all livestock, they came to an overall figure of 7%–8%—that is, roughly, a ratio of 1:13, for both calories and protein.[23] As with other studies, cattle were by far the least efficient feed converters—returning in a 96%–97% loss, or a feed conversion ratio of, roughly, 1:24. These figures are consistent with other mid-range estimates, and it is useful to have both calories and protein being used in the assessment. The case I am building for abandoning the eating of animals is based on analysis of energy, specifically EROI. Therefore, you might assume that the calorie-based ratios are more pertinent to my goals. That would be misleading. Both calories and protein are important because both are, in different ways, indices of energy. Protein synthesis is the basis of any biological construction. With it, you can build a certain number of cells, and the mass of these cells will be equivalent to a certain amount of energy. In the end, energy is what counts. But protein is, in this sense, one way of measuring energy. All things are an exchange for fire, and fire for all things.

What interests me about the Shepon paper are not the figures—which are consonant with many other studies—but the moral they draw from these figures. Like the *Chick Fil-A* advertising campaign, initiated approximately 20 years ago but still going strong today, they decided the moral of the story was to *Eat Mor Chikin*. Imagine. You have been investing in the stock market as long as you can remember, and, quite frankly, have lost an awful amount of

---

[22] The distinction parallels that between various types of EROI, in particular, the distinction between mine-mouth EROI and point-of-use EROI. In the interests of fairness, I should point out that we would also have to subtract from the plant matter weight all those parts of the plant animals eat but that we don't or can't.

[23] A. Shepon, G. Eshel, E. Noor, and R. Milo, "Energy and Protein Feed to Food Conversion Efficiencies in the US and Potential Food Security Gains from Dietary Changes," *Environmental Research Letters*, September 2016, https://iopscience.iop.org/article/10.1088/1748-9326/11/10/105002/pdf.

money. You thought it was simply because you are a financial moron, just like your daddy, and his daddy before him. But, in fact, your traders have been skimming vast amounts off your trades, even before they were made. Some of the traders were admittedly worse than others. Mr. Beef was the worst. Every time you gave him $100 dollars to invest, he pocketed $96 and invested only the remaining $4. Mr. Bacon was slightly better; he would only skim around $90 off the top. Mr. Chikin was the best—although, let's face it, the bar hasn't been set very high—skimming merely a little over $80 off your $100 investment.[24] Finally, you discover what has been happening. What conclusion would you draw from this? More importantly, what conclusion *should* you draw? If your conclusion is, "Oh well, I guess I should let Mr. Chikin handle my trades from now on," then you would indeed be a financial moron. A better conclusion would, of course, be: stop making trades in such criminally disadvantageous circumstances. Especially if you don't have to make these trades.

Breathing Space

We do not need to enter into the kinds of warped energy exchanges that are definitive of the meat industry—exchanges that, from the perspective of their EROIs, are about as upside-down as my mortgage will be when people finally work out that Miami is going under. It is possible for every single one of us to give up eating meat today. Right now, in fact. When you get to the end of this sentence. Now. It might not be easy. And, for many of us, it might not be pleasant. But it is possible. Nor would a particularly lengthy transition or phasing-out period be required. The average lifespan of a broiler chicken is 6 weeks. That of a factory farmed pig is 5 months.

As I argued earlier, it would be a mistake to think we can separate the issue of GHG emissions from the issue of how we acquire our energy. For complex structures, such as humans and our societies, it's all about energy coming in versus energy going out. The GHGs that result from the meat industry are the result of a certain strategy that we have adopted for acquiring energy. The GHG emissions that result from the industry can't be separated from the EROI of the strategy. Energy is required for the plants we grow to

---

[24] Figures employed are (roughly) those of Shepon et al., "Energy and Protein Feed to Food Conversion Efficiencies in the US and Potential Food Security Gains from Dietary Changes."

feed livestock. Feed requires energy to produce: energy involved in planting seeds, fertilizer for them to grow, harvesting, storage, transport. We then give this feed to animals, and most of this energy that we have invested into the feed is lost. The meat industry is, fundamentally, an energy exchange, one that has gone badly awry, and the GHGs produced by this industry are completely avoidable.

If we abandoned this exchange, changed our energy investment portfolio, then we would, very quickly, stop pumping somewhere in the region 14.5% of the world's GHGs into our atmosphere. The 7.1 gigatons of CO_2-equivalent gases that this industry spews into the atmosphere every year would largely cease. This would give us something whose value cannot, in our current predicament, be overestimated. Breathing space. Put this figure of 14.5% in the context of a set of renewable energy sources that are necessary but hovering on the brink of viability—perched unsteadily on or near the edge of the net energy cliff. Suppose now that we could somehow, magically, reduce the amount of GHGs entering the atmosphere by 14.5%. This might be the crucial difference that changes everything. It gives us the time to find ways of developing the, still relatively new, solar, wind, and wave technologies and pushing these sources back from the edge of the cliff.

Breathing. Space. Free—that is, nonsequestered—carbon dioxide has a half-life in the atmosphere of 50–100 years, but much of it sticks around long after that. The carbon dioxide that we have already put into the atmosphere isn't going anywhere anytime soon. That's why a certain amount of global temperature rise is already, as it is said, *locked in*.[25] The more carbon dioxide we put into the atmosphere, the greater is the locked-in temperature increase. But there are certain things we can do to mitigate this. Carbon stays around in the atmosphere for a long time, and as such, it is adept at locking in future temperature increases. Methane, as we have seen, is much better at trapping heat in the short to medium term—over a 20-year time frame, methane is 72 times better at trapping heat than carbon dioxide, and 25 times better over a 100-year time frame. However, against this, its atmospheric half-life is only 8.6 years. It will disappear from the atmosphere very quickly. This means that it is nowhere near as good as carbon dioxide at locking in global temperature increases.

---

[25] Even in 2014, the World Bank estimated 1.5°C of warming was already locked in: https://www.worldbank.org/en/news/feature/2014/11/23/climate-report-finds-temperature-rise-locked-in-risks-rising. The International Panel on Climate Change (IPCC) is a little, but only a little, more upbeat. See footnote 28.

Given this fact, recall the FAO report, commissioned by the United Nations.[26] According to this, roughly 44% of livestock emissions take the form of methane. That is, 44% of 7.1 gigatons of CO_2-equivalent GHGs from the meat industry is made up of methane. Equivalently, methane comprises 3.1 gigatons of the CO_2-equivalent GHG emissions that derive from the meat industry.[27] Thus, at the very least, we can conclude that vast quantities of CO_2-equivalent GHG emissions can be removed from the atmosphere *within the next decade*. If we stop eating meat. The United Nations tells us that we have to act quickly. We need to reduce total GHG emissions by 45% by 2030 if we are to stay within the 1.5°C threshold of global temperature increase.[28] Within this time frame, the most constructive thing we can do is take methane out of the atmosphere, and we can do this, in large quantities, if we no longer eat meat. Of course, it is crucial not to let carbon dioxide emissions get away from us any further: that will simply result in further locked-in warming. But by taking this amount of atmospheric methane out of the equation and removing the upside-down EROIs of animal flesh and products, we will give nascent renewable technologies the breathing space they need to— hopefully—move a few steps away from the edge of the net energy cliff. If they can do this, then the removal of anthropogenic carbon dioxide from the atmosphere will take care of itself in the coming decades. Taking methane out of the atmosphere will buy us this time. And we take methane out of the atmosphere through the expedient of no longer eating meat.

There is more argument to come. In fact, as things stand, we have only looked at half of the climate-based case against the meat industry. Giving up meat not only significantly reduces GHG emissions, including those all-important (in the short term at least) methane emissions. It also allows us to do something else: something that, in the struggle against climate change, is even more important. This is the subject of the next chapter.

---

[26] *Tackling Climate Change through Livestock: A Global Assessment of Emissions and Mitigation Opportunities* (2013).

[27] Note also that this figure of 3.1 gigatons is the figure provided by the FAO. It has been calculated on the basis of a 100-year, rather than 20-year, time frame, where methane has only a GWP of 28 rather than the 72 GWP it possesses over 20 years.

[28] See International Panel on Climate Change, Special Report (2018), *Global Warming of 1.5°C*, https://www.ipcc.ch/sr15/.

7

A Forest Future?

Natural Carbon Capture and Sequestration

In Chapter 5, I argued that technological carbon capture and sequestration (CCS) devices are, certainly when assessed in terms of what they will do to the energy returned on energy invested (EROI) of a fossil energy source, unpromising. Moreover, given the simple physical constraints involved in what they are supposed to do—the ratio of source to effluent, and the difficulties involved in storing, transporting, and disposing of the latter—they may well be destined to remain so. But the best CCS devices in the world were built not by us but by nature. We know them as *trees*.

Trees, and other plants, remove carbon dioxide from the atmosphere by photosynthesis. Through photosynthesis, driven by energy supplied by the sun, plants produce carbohydrates. They do so by fusing the carbon contained in atmospheric carbon dioxide with hydrogen and oxygen. These carbohydrates are then used in the growth and maintenance of the plant, and any carbohydrates not used for this purpose are moved to the soil through its roots, where they are fed to dirt-bound fungi, which in turn feed minerals back to the plant. Atmospheric carbon is, thereby, *sequestered*: moved from the air to both the plant and the soil where it is stored. The bigger the plant, the more carbon is stored in it. And the deeper its roots, the better it is at sequestering carbon in the soil. Both of these considerations point to trees being the best of the natural sequestration devices.

Carbon is stored in trees—as we shall see, roughly 50% of the non-water weight (or *dry weight*) of trees is made up of carbon—and in soil. Both forms of storage have their advantages and disadvantages. Soil is, in general, a much more stable, robust carbon *sink*—that is, carbon store. Trees can suffer misadventures—through fire, logging, parasites, and so on, and when they die, their carbon is released back into the atmosphere. The same is not true of soil. Although carbon can be released from soil through degradation, this is generally a slower process than the death and decomposition of a tree. The disadvantage of soil, however, is that it eventually becomes saturated

World on Fire. Mark Rowlands, Oxford University Press. © Oxford University Press 2021.
DOI: 10.1093/oso/9780197541890.003.0007

with carbon, and after this point it will be able to absorb no more. Thus, the storage limits of carbon in the soil are strictly finite, and in mature woodland soil they will have typically reached the carbon saturation point.

In this chapter, I shall further the climate-based case against eating animals. The central idea for which I shall argue is that abandoning this habit will make available vast areas of land on which we can then do perhaps the most important thing we can do in the fight to ward off the climate crisis: plant trees. Trees, I shall argue, are of crucial importance in this fight. We simply need to find a way of planting enough of them.

Lockdown Fun

Here's something to do when on lockdown. Something that I like to do when I want my kids to stop playing *GTA* for a while. Or *Fortnite*. Or FIFA19. Or when I just want to annoy them. SCIENCE PROJECT!! Today, boys, we are going to calculate the amount of carbon dioxide the trees in our back yard sequester every year. "No!" "Why!" "Every tree!" And so on, and so forth. These protests will continue for some time. Nevertheless, we are doing it. I won't actually make them do every tree—I'm just winding them up there. A representative sample will do very nicely. Nor is mathematical precision our goal. In this chapter, I am interested in painting a picture of the enormous potential of what is known as *afforestation*: the introduction of trees to areas that have not been forested in the recent past. It is a picture, painted in broad strokes, that I seek, rather than a precisely calculated mathematical proof. Therefore, for the purpose of our little garden exercise this morning, ballpark figures will do just nicely.

We begin with the southern live oak tree in the backyard. This tree used to annoy me, because of its habit of dropping leaves in the pool and clogging up the filters. But that was before I saw the light. The first stage is to determine what's known as the total *green weight* of the tree.[1] To do this, you first need the tree's diameter. This you can work out from the old formula, learned and then subsequently forgotten, by every schoolchild:

---

[1] Calculation method based on A. Clark, J. Saucier, and W. MacNab, "Total Tree Weight, Stem Weight and Volume Tables for Hardwood Trees in the Southeast," *Georgia Forest Research Paper* 60 (1986): 45–50. See also http://www.unm.edu/~jbrink/365/Documents/Calculating_tree_carbon.pdf.

$$C = \pi D \text{ (or } C = 2\pi r)$$

The circumference is equal to the diameter (= 2× the radius, r) multiplied by π. It follows from this that the diameter is the circumference divided by π. So we measure the circumference of the tree, around 4 feet from the ground. As I mentioned, precision is not the goal in any of this. At each stage, we are just looking for ballpark figures. And, in accordance with a general strategy pursued through this book, I'll round down so as not to be accused of cheating. The circumference turns out to be just over 70 inches. Divided by 3.14—the value of π—gives us a diameter of roughly 22 inches. From this we can calculate the green weight of the tree, using the formula:

$$W = 0.25D^2H$$

W = the green weight of the tree. D = the diameter of the tree. H = the tree's height. The green weight of the tree (in pounds) is one-quarter multiplied by the square of the tree's diameter (in inches) multiplied by the tree's height (in feet). So now we need the height. How do you get this? Well, you can, if you like, be thorough. Use a protractor to measure the angle from the horizontal to the top of the tree. Call this angle A. Then measure the distance to the foot of the tree. Call this distance b. Then, finally, measure the height of the observer's eyes—that is, the one who held the protractor—above the ground. Call this height z. Then, the height of the tree can be calculated using the following formula:

$$H = (\tan A \times b) + z$$

You could do that. Though I have to admit that is a little hard core even for me, and it would result in open insurrection on the part of my sons. I have a simpler method: educated guessing. Whenever a hurricane is about to come through South Florida, I have to spend an uncomfortable couple of days or so up a 24 foot ladder, putting up the shutters. It's not a lot of fun, especially as I'm not a big fan of heights. It looks a lot higher when you are at the top of the ladder looking down than when you are the bottom of the ladder looking up, believe me. The height of the roof at its apex, I would estimate, based on my intimate knowledge of 24 feet, is at least 30 feet, and probably a little more than that. And the last time I was on that roof—inspecting possible damage after Irma in September 2017—I would estimate that the tree was at least 10 feet taller than the house. So the tree is probably somewhere in the region of 40 feet tall, or maybe a little taller. But, in the spirit of rounding down, and to guard against any mistakes or excessive optimism on my part, let's suppose it is only 35 feet tall. Armed with these figures, we can now calculate the green weight of the tree using the earlier formula:

$$W = 0.25(22^2)(35)$$
$$W = 0.25 \times 484 \times 35$$
$$= 4235 \text{ pounds}$$

Let's say 4200 pounds. However, the tree is more than just trunk and branches. There are also the roots, which are obvious since they have come to completely dominate the south quadrant of our garden. The general consensus is that the root system of a tree will add around 20% to the tree's green weight.[2] This equals 840 pounds. Therefore, the total green weight of the tree is:

$$\text{Total Green Weight, } W = 4200 + 840 \text{ pounds}$$
$$= 5040 \text{ pounds}$$

The next step is to calculate what's known as the *dry weight* of the tree. This is the green weight of the tree minus the weight of the moisture contained in it. Trees in general contain a lot of moisture, and the green weights of trees usually comprise around 50% moisture weight. Southern live oaks, however, are wetter than most, with moisture content generally estimated to be in the 65%–75% range. Let's assume 70%.[3] This means that the dry weight of the tree will be roughly 30% of its green weight. Therefore:

$$\text{Dry weight} = 1512 \text{ pounds}$$

[2] This is a conservative estimate. For example, M. Cairns, S. Brown, E. Helmer, and G. Baumgardner, "Root Biomass Allocation in the World's Upland Forests," *Oecologia* 111 (1997): 1–11, estimate an additional 26% of the tree's green weight is provided by roots.

[3] Figures obtained from the Warnell School of Forestry, University of Georgia, https://www.warnell.uga.edu/sites/default/files/publications/Live%20oak%20SEH%20%26%20ID%20pub_15-6.pdf.

The next step is to calculate the amount of carbon sequestered in this dry weight. This is roughly 50% of the tree's dry weight:[4]

$$\text{Carbon sequestered} = 1512 \div 2$$
$$= 756 \text{ pounds}$$

The tree, that is, has sequestered approximately 750 pounds of carbon during the course of its life. How long has this life been? This is where things get a little hazy, but I can, again, make some educated guesses. The neighbors tell me—and Zillow confirms—that the house was built in 1989, a mere 3 years before Hurricane Andrew. Maybe a young oak tree would have survived Andrew, or maybe it was planted shortly after. Either way, today the tree is likely to be in the vicinity of 30 years of age. If this is correct, then to calculate the amount of carbon sequestered per year, we can simply divide the total amount of carbon sequestered by 30:

$$756 \div 30 = 25.2 \text{ pounds}$$

Let's say 25. Carbon is not the same as carbon dioxide, of course. Carbon has an atomic weight of 12, oxygen has a weight of 16 (both roughly). Since carbon dioxide is made up of one carbon atom combined with two oxygen atoms, the atomic weight of carbon dioxide is 44. Therefore, every unit of carbon sequestered is equivalent to 44/12 units of carbon dioxide = (roughly) 3.6 units of carbon dioxide. To calculate the total amount of carbon dioxide sequestered by the tree, therefore, we can multiply the total amount of carbon sequestered by 3.6:

$$\text{Total Carbon Dioxide sequestered by tree} = 756 \times 3.6$$
$$= 2721.6 \text{ pounds}$$

Let's say 2700. If so:

$$\text{Amount of carbon dioxide sequestered per year} = 2700 \div 30$$
$$= 90 \text{ pounds}$$

---

[4] Figure obtained from *Natural Resources Wales*, https://cdn.naturalresources.wales/media/687190/eng-worksheet-carbon-storage-calculator.pdf.

These are, obviously, amateurish calculations and, for that reason, I am going to rest absolutely nothing on them in the argument to follow, I assure you. In fact, however, the figure of 90 pounds of carbon dioxide sequestered per year is not *that* far away from estimates made by at least *some* of the professionals. Admittedly, it is quite far away from others of those estimates. For example, according to North Carolina State University: "A tree can absorb as much as 48 pounds of carbon dioxide per year and can sequester 1 ton of carbon dioxide by the time it reaches 40 years old."[5] If correct, my oak tree has been putting up some suspiciously high numbers, both in terms of total amount of carbon dioxide sequestered and the rate of sequestration per year. If sequestering carbon dioxide were an Olympic event, you would have to suspect my tree was doping. More likely, however, is that it is more akin to a testing error—my amateurish calculations are awry. More encouraging, however, are the estimates of Nowak and colleagues, U.S. Forest Service scientists at the Center for Urban Forest Research, who place the figure at 88 pounds of carbon dioxide per year.[6] The performance of my southern live oak would still be a little high but almost within normal parameters—perhaps more Usain Bolt than Justin Gatlin.

One should also bear in mind that the southern live oaks—there are three more just like it in the front garden—are, by some margin, the biggest trees on our property. It would be interesting to see how the figures look for some of the smaller trees. Thus—just when they think they are going to get back to *Fortnite*—I force my sons to do the same thing with another tree, one of the queen palms at the front of the house. At a height of 4 feet off the ground, the diameter of this tree is approximately 11 inches. That's a tape measure estimate—we're under the *Fortnite* gun and have dispensed with $C = 2\pi r$. The height is easy to estimate: it is slightly taller than the house—c. 30 feet. Therefore:

$$\text{Green weight} = 0.25 \times D^2 \times H$$
$$= 0.25 \times 11^2 \times 30$$
$$= 907 \text{ pounds}$$

[5] https://projects.ncsu.edu/project/treesofstrength/treefact.htm. Accessed June 9, 2019.

[6] D. Nowak, E. Greenfield, R. Hoehn, and E. Lapoint, "Carbon Storage and Sequestration by Trees in Urban and Community Areas of the United States," *Environmental Pollution* 178 (2013): 229–236, https://www.fs.fed.us/nrs/pubs/jrnl/2013/nrs_2013_nowak_001.pdf.

Add 20% for the roots:

Total Green Weight = 1088 pounds

The dry weight of the tree is equal to the green weight minus the water weight. Palm trees, in general, are not as thirsty as live oaks, and their percentage of water weight is, correspondingly, less: somewhere in the region of 50% seems to be the consensus. Therefore:

Dry Weight = 544 pounds

The amount of carbon contained in the tree is this figure divided by two:

Carbon contained in tree = 272 pounds

Carbon dioxide sequestered by the tree is, therefore, 272 multiplied by 3.6 (i.e., 44/12):

Total Carbon Dioxide sequestered = 979 pounds

Divided by the 30 years I estimate the tree to have been alive:

Carbon Dioxide sequestered per year = 32.6 pounds

Let's say 32 pounds—certainly consonant with the "as much as" estimations of North Carolina. And while it is, perhaps, a little disappointingly low with regard to the U.S. Forest Service figures—I shall have to have a word with my queen palms to ensure they are putting forth maximum effort—it is certainly within reasonable parameters.

My garden is largely a mixture of southern live oaks (4), queen palms (3), royal palms (3), and coconut palms (9). To round up the numbers, and thus make calculations simpler, I'm also going to lay claim to a Poinciana, which straddles the boundary between my property and my neighbor's. It's not very tall, but it has an immensely thick trunk at 4 feet off the ground, and I find I simply can't resist the D^2. This gives us 20 trees, with varying amounts of carbon-sequestering capacities. How much store should we place in my amateurish calculations? Absolutely none, of course. I can't emphasize that enough. But it does not seem unreasonable to use the figures of North Carolina and the U.S. Forest Service as the basis for calculations. North Carolina says as much as 48 pounds per year. U.S. Forest Service tells us 88 pounds is doable. Somewhere in the region of the North Carolina figures would be a conservative choice. So let's assume 50 pounds of carbon dioxide per year, per tree. If so, the trees on our property should sequester the pleasantly even figure of 1000 pounds of carbon dioxide a year. Roughly, of course. This is a little under half a metric ton of carbon dioxide (= 2204 pounds). How significant is this?

A Bigger Garden

In the great scheme of things, a little under half a metric ton—give or take—of carbon dioxide sequestering per year may seem utterly insignificant. According to the World Bank, the average American is responsible for 16.5 metric tons of carbon dioxide per year.[7] So, between the four of us and the dog, there is going to be somewhere in the region of 66 metric tons of carbon dioxide that need to be offset. Prima facie, 65.5 metric tons seems to be a negligible improvement. If only we had a bigger garden. Yes, a bigger garden: that, really, is the key. The Rowlands family residence sits, according to Zillow, on 0.38 acres. An acre is equal to 43,560 square feet. Multiplying that by 0.38 gives us 16,552 square feet. Of that, the house is approximately 3500 square feet. Subtracting that gives us 13,052 square feet. This area of 13,052 square feet currently contains 20 trees. That is a tree roughly every 650 square feet. But it could easily contain quite a few more. There are treeless lawn areas out the front and back, a driveway, a patio, and a pool. Surveying the relative size of these compared to the land occupied by trees, it is not unreasonable to suppose that this area could accommodate at least twice the number of trees it presently harbors without overcrowding. And I suspect this is a conservative estimate. That would be a tree every 325 square feet, if we assume a mixed assemblage of different species as is currently the case in our garden. Dividing an acre, that is, 43,560 square feet by 325 gives us 134 trees per acre, of mixed varieties of trees, some tall with relatively narrow foliage (such as the queen, royal, and coconut palms) and others that like to spread (such as the oaks and the Poinciana).

---

[7] World Bank, https://data.worldbank.org/indicator/EN.ATM.CO2E.PC?locations=US. On various websites, you will often see the figure of 4 metric tons or so. That figure is for carbon rather than carbon dioxide. One unit of carbon is equal to 3.67 units of carbon dioxide.

Rather gratifyingly, this number is also reasonably consistent with those who do this sort of thing for a living. According to the University of Maryland extension bulletin EB-407 (2013), 150 trees per acre is a reasonable number for mature trees.[8] My number is a little lower, in all likelihood because the trees in my estimates are of mixed varieties rather than monoculture. I would happily go with the professional estimate of 150 trees per acre over my amateurish calculations. But in the interests of conservatism, and to avoid being accused of cheating, I'll go with my lower figure—but I'll bump it up from 134 to 135. That only seems fair.

If the 20 trees on my little parcel of land sequester roughly 1000 pounds of carbon dioxide per year, then 135 trees would sequester roughly 6750 pounds—around 3 metric tons (3.06 to be precise, but let's say 3) of carbon dioxide per year. How reliable is this figure—based, as it ultimately is, on an attempt to get my sons away from Xbox for an hour or two? Gratifyingly—by some miracle or other—it is consistent with the kinds of figures presented by those who know what they are doing. The U.S. Environmental Protection Agency (EPA) estimates that *afforestation*—putting trees on land where there were no trees in the recent past—would result in the sequestration of carbon dioxide at a rate of between 2.2 and 9.5 metric tons per acre per year.[9] The estimates of the U.S. Department of Agriculture (USDA), however, range from 2.7 to 7.7 metric tons of carbon dioxide sequestered per acre per year.[10] My figure of 3 metric tons lies at the conservative end of both of these estimate ranges, and this is unsurprising given the conservative estimates I have adopted in the calculations so far. I have been conservative on root weight, conservative on amount of carbon dioxide sequestered per tree per year, and conservative on number of trees per acre. This has yielded an estimate of 3 metric tons per acre per year of carbon dioxide sequestered. But there's conservative, and then there is pathologically conservative. The EPA

[8] According to the University of Maryland extension bulletin EB407 2013, "There may be well over 10,000 seedlings per acre in the first 5 years. This number decreases to about 1,000 as the stand matures to a point when the trees average 5–11 inches DBH (pole-sized). Most of the young trees naturally die as other trees out-compete them for sunlight. When trees average more than 11 inches DBH (saw-timber size), the number of trees declines to 500 trees per acre, and eventually to 150 trees per acre in very mature woodlands." http://4hforestryinvitational.org/training/forest-evaluation-training/EB-407_ForestThinning.pdf.

[9] U.S. Environmental Protection Agency, Office of Atmospheric Programs, Greenhouse Gas Mitigation Potential in U.S. Forestry and Agriculture, EPA 430-R-05-006, Washington, DC, November 2005, Table 2-1, http://www.epa.gov/sequestration/pdf/greenhousegas2005.pdf.

[10] Jan Lewandrowski, Mark Peters, and Carol Jones et al., "Economics of Sequestering Carbon in the U.S. Agricultural Sector," USDA Economic Research Service, Technical Bulletin TB-1909, Washington, DC, April 2004, Table 2.2, http://www.ers.usda.gov/ Publications/TB1909/.

calculates between 2.2 and 9.5, and the USDA tells us between 2.7 and 7.7 metric tons of carbon dioxide sequestered per acre per year. Without, in any way, abandoning my strategic methodological conservatism, I think I can legitimately assume a figure of 4 metric tons per year of carbon dioxide sequestered by an acre of forest per year. So I'm going to live a little and do just that. It still places me, more or less, at the conservative end of the range of calculations provided by the EPA and USDA.

If we assume that an acre of trees will sequester carbon dioxide at a rate of 4 metric tons per year, then to offset the 66 metrics of carbon dioxide the Rowlands family produces every year would, therefore, require (66 divided by 4) roughly 16.5 acres. If our garden were 16.5 acres, and devoted entirely to growing trees, we would have achieved carbon neutrality. That's 4.125 acres per person, on average. There's not much chance of getting a garden that size in Miami—unless this book sells a *lot* better than I anticipate. But it's good to know. And there is more good news on the way.

Don't Forget the Dirt

The carbon stored in the trees is not the end of our garden's arboreal attempt to combat climate change. As I mentioned at the beginning of this chapter, carbon storage—the result of the sequestration efforts of my trees—can take place both in the trees themselves and also in the soil in which the trees are rooted. And as Paustian and colleagues point out, "Unmanaged forests and grasslands typically allocate a large fraction of their biomass production belowground and their soils are relatively undisturbed."[11] To properly estimate the amount of carbon sequestered in our yard, therefore, we need to address the carbon stored in the soil as well as in the trees. What sort of ratio of storage-in-trees to storage-in-soil might we expect?

The study of the carbon-storage ability of soils is, in many ways, in its infancy, and so specific figures are not as widely available as one would like. However, one way we might approach this is to look at loss of carbon in soils that have been transformed from forest to cropland. The advantage of this method is that there is, in fact, reasonably extensive documentation of the kind of loss involved. Estimated losses typically range between 0.5 to >2

---

[11] See, for example, K. Paustian, J. Lehmann, S. Ogle, D. Reay, G. Roberstson, and P. Smith, "Climate-Smart Soils," *Nature*, April 7, 2016, 49–57.

metric tons (or "megagrams") of carbon per hectare per year when such conversion takes place.[12] This equates, roughly to between 0.2 and 0.8 metric tons per acre per year. This is carbon loss. To calculate the amount of carbon dioxide sequestration this corresponds to, we multiply, as earlier, by 3.6 (44/12). This gives us figures of between 0.73 and 2.9 metric tons of carbon dioxide sequestration that are lost per acre per year in the conversion of forest into cropland. This, therefore, is the difference between the amount of carbon dioxide sequestered in forest soil versus grazing or cropland soil. Let's assume a neat figure on the conservative end of this spectrum—as I say, I don't want to be accused of stacking the deck in my favor. Let's say 1 metric ton per acre per year.

When we took just the trees into account, we arrived at an estimate of 4 metric tons of carbon dioxide per acre per year. To offset the 66 metric tons of carbon dioxide produced annually by the Rowlands family, we would need (66 ÷ 4 =) 16.5 acres of woodland, which equates to 4.125 acres per person. Once we take the soil into account, however, the amount of carbon dioxide sequestered per acre per year turns out to be at least 5 metric tons. Thus, to offset the 66 metric tons we produce, we would now need (66 ÷ 5 =) 13.2 acres of acres of woodland. This would amount to 3.3 acres per person. If only we could find 3.3 acres of land to afforest, per person, we could offset the carbon footprint of the most energy-intensive nation in the history of the world. Maybe.

We are talking about what's known as *afforestation* rather than *reforestation*. Reforestation is the replanting of trees on land that has recently been used to grow trees, as when trees are replanted on land that has recently been logged, for example. Afforestation, however, occurs when trees are replanted on land that has not recently contained trees—as when, for example, we plant trees on land recently dedicated to crops or pasture. The difference is that in recently or currently forested land, soil will likely have already reached the carbon saturation point. This would result in having to reduce the amount of carbon sequestered per acre by roughly 20%—circa 4 tons per acre instead of 5 tons. So let me be clear that I am discussing afforestation here.

Where could we possibly find this new land to afforest? This is where our old habit of eating meat re-enters the picture. Abandoning the habit of eating

[12] E. Davidson and L. Ackerman, "Changes in Soil Carbon Inventories Following Cultivation of Previously Untilled Soils," *Biogeochemistry* 20 (1993): 161–193. S. Ogle, F. Breidt, and K. Paustian, "Agricultural Management Impacts on Soil Organic Carbon Storage under Moist and Dry Climatic Conditions of Temperate and Tropical Regions," *Biogeochemistry* 72 (2005): 87–121.

meat may not take us all the way to where we need to go. But, I shall argue, it will take us a considerable distance in the right direction. Indeed, it might take us almost all the way there. Like my sons, now happily back to playing *GTA*, it will at least get us *back in the game*. The goal of the game is to get to the magic figure of 3.3 new acres of woodland per person. The current population of the United States is 331 million, give or take. This means we would need to find somewhere in the region of 1,092,000,000 acres—1.092 billion—new acres of forest to achieve carbon neutrality for the United States. Let's say 1.1. That does seem like a lot. But if we stop eating meat, then we have a chance of getting there—especially, as I shall try to show—what counts as "there" is rapidly evolving. That's a little cryptic, I know. We'll get there, eventually.

Gaining Ground

In the contiguous United States, roughly 41% of land is used for grazing and for growing crops for livestock feed. This is far more than any other kind of land use. According to the USDA, in 2012 (the most recent year for which we have figures) there was the following distribution of cropland in the United States.[13]

- Total cropland = 392 million acres
- 52 million acres lay idle or fallow—recovering for future use.
- 77.3 million acres was dedicated to growing crops for human consumption.
- 127.4 million acres were used for growing crops—principally corn and soy—to make livestock feed.

Thus, in the United States, roughly 1.7 times as much land is used to grow crops for livestock consumption as is used to grow crops for human consumption. If we afforested this land, we would already have made a dent in the 1.1 billion acres of new forest required to offset the annual national carbon footprint. It is true that we would have to grow more crops for human consumption. But remember the feed-conversion ratios we encountered in

---

[13] https://www.ers.usda.gov/topics/farm-economy/land-use-land-value-tenure/major-land-uses/.

Chapter 6: anywhere between 1:4 and 1:25 (or higher) on some estimates. Thus, depending on the animals for which we are growing the crops, we would, roughly, need to dedicate anywhere between one-quarter and 1/25th as much land to growing replacement crops for us as we currently dedicate to growing crops for animals. One quarter of 127 million acres is 31.85 million acres. One-twenty-fifth of 127.4 million acres is 5.1 million acres. Therefore, it should be clear that I am doing myself no favors by allowing an extra 27.4 million acres to be added to the land we would employ for growing crops should we abandon the meat industry. It would probably be considerably less than this. This gives us a nice round saving of 100 million acres. But it doesn't stop there. According to the USDA:

- 38.1 million acres of cropland is used to grow corn for ethanol.

We have already seen what a disaster this is in EROI based-terms—an EROI of 1–2 is simply not a viable way of producing energy. Perversely—and this is, of course, a perverse age—the USDA also tells us that the demand for bioethanol is the most significant current factor in driving up the price of corn.[14] Bioethanol and biodiesel are not, speaking EROI-cally, even in the game. We should, therefore, reassign the use of these 38.1 million acres. Let's add that to our 100 million acres of land saved already. We now have 138 million acres of land available for afforestation. There is still, however, further land available:

- 62.8 million acres used for growing what the USDA categorize as *other grain or feed exports*.

Not all grain need be used for animal feed, but most is. The USDA distinguishes this category from that of *wheat exports*, which is almost entirely used for human consumption. This suggests most of this category consists in grain exports used in animal feed. In the United States itself, the ratio of grain used in animal feed to that used to feed humans is 1.7 to 1. But let's be conservative and assume only two-thirds of this category is devoted to growing grains for animal feed. Dispensing with this would supply us with,

---

[14] As the USDA put it: "Strong demand for ethanol production has resulted in higher corn prices and has provided incentives for farmers to increase corn acreage." https://www.ers.usda.gov/webdocs/publications/84880/eib-178.pdf?v=6831.8.

roughly, an extra 42 million acres. Thus, our total of land newly available for afforestation rises to 180 million acres.

By far the largest category of land use in the United States—outstripping all other uses by considerable margins—is *grazing*. According to the USDA:

- 654.1 million acres of land are used for the grazing of livestock.

Some of these acres lie on private land, and the rest lies on public lands which can be grazed for a fee. If we added these acres to the ones reforested, this would give us an overall acreage of roughly 834 million acres newly available for reforestation. This is roughly three-quarters of the acreage required to offset the annual carbon footprint of the nation. That is, 1.1 billion minus 834 million leaves us with 266 million acres. That is the current shortfall: the additional acreage we need to find in order to offset the carbon footprint of the United States. It's a lot, but it's a lot less than 1.1 billion.

We can make the same point at the level of the individual. We have calculated that 3.3 is the magic number: the amount of acreage per person we would need to find to offset the carbon footprint of the average resident of the United States. This assumes the World Bank estimate that the average US resident is responsible for 16.5 metric tons of carbon dioxide per year. If 834 million of the 1.1 billion required acres is offset through reforestation, this leaves us with 266 million acres divided by the population of 331 million. This means we are roughly 0.8 acres short, per person, of carbon neutrality. The acquisition of 834 million acres, gained through abandoning the meat industry, has bought us 2.5 acres person. We are not there yet. But we are getting there. Slightly more than 77% of the way there, in fact. And there is one final twist still to come.

The Changing Thereabouts of There

The final change turns on the earlier allusion of the evolving location of "there." "There" may, in fact, no longer be where we thought it was. In calculating the annual national carbon debt of the United States, I used the figures of the World Bank that estimates the average carbon footprint of each person at 16.5 metric tons. But these calculations assumed a functioning meat industry that, according to the FAO, produces 7.1 gigatons of CO_2-equivalent gases each year. Part of the carbon footprint of the average

American—whether or not they eat meat—is a function of the amount of the quantity of emissions produced by this industry. How much of the 7.1 gigatons is produced by the United States? According to a study commissioned by the EPA, agriculture in general produces 574 million metric tons of CO_2-equivalent gases.[15] Of this amount, 42% (or 214 million tons) comes from the meat industry directly. This figure does not include emissions from crops grown to produce animal feed, which, at present, stand in a ratio of 1.7:1 vis-à-vis crops grown to produce food for humans. So, subtracting 214 from 574 million, we arrive at 360 million tons and can expect at least two-thirds of this to be emissions resulting from crops grown to produce animal feed or biofuels. This amounts to 240 million tons. These, I hasten to point out, are likely to be conservative estimates. Thus, we should expect that of the 574 million tons of CO_2-equivalent gases produced by the US agricultural industry, a total of (214 + 240) 351 million tons of CO_2-equivalent gases derive annually from specifically animal agriculture. That is, 351 million divided by the 327 million population of the United States equals 1.07. Originally, we assumed, with the World Bank, that the average carbon footprint of a US citizen was 16.5 metric tons. But, with the abandonment of the meat industry advocated in the previous chapter, that would be reduced to around 15.4 metric tons.

In making these estimates, we have to be aware of the possibility of *double counting*. The reason this is a possibility is that changes in land use have already been factored into the United Nations estimates of the CO_2-equivalent gases produced by the global meat industry. In touting the greenhouse gas (GHG)-mitigating effects of changes in land use, it may seem as if I am counting the same emissions twice. This, in fact, is not so since there is a crucial difference in the method of counting. In the United Nations estimations of CO_2-equivalent gases produced by the meat industry, changes in land use count toward such emissions only once—essentially, in the year that the changes occur.[16] However, when we talk of the GHG mitigation afforded by afforestation, this mitigation occurs not only in the year afforestation begins (or, if you prefer, when afforested areas attain a certain level of maturity) but also every year thereafter—every year in which the new forests continue to

---

[15] https://19january2017snapshot.epa.gov/sites/production/files/2016-04/documents/us-ghg-inventory-2016-chapter-5-agriculture.pdf.

[16] United Nations, Food and Agriculture Organization, *Tackling Climate Change through Livestock: A Global Assessment of Emissions and Mitigation Opportunities*, http://www.fao.org/3/a-i3437e.pdf.

thrive.[17] If there is a worry about double counting here at all, it would apply only to the first year of reforestation, and not to the (hopefully) hundreds or thousands of subsequent years. Therefore, the effects of double counting can, for all practical purposes, be ignored. In the absence of a meat industry, the carbon footprint of the average American would be 15.4 metric tons rather than 16.5 metric tons.

This may not sound like much of a difference, but small margins can make a large difference. Think first of the annual carbon debt of the Rowlands family. Formerly calculated at 66 metric tons per year, it now reduces to 61.6 metric tons, not including the footprint of our dog Shadow. If we assume, again, a total of 5 metric tons of carbon dioxide sequestration per acre per year, then instead of needing (66 ÷ 5=) 16.5 acres of acres of woodland, or 4.125 acres person, we would only need (61.6 ÷ 5=) 12.3 acres, which would amount to 3.07 acres per person. Remember, however, that we have already around 2.5 acres of that through the afforestation afforded by abandoning the meat industry. We are short by only 0.57 acres of forest per person. Essentially, through the expediency of no longer eating meat and reforesting the land formerly allocated to the meat industry, the most energy-hungry society the world has ever known would be roughly 80% of the way to carbon neutrality.

Some Competing Factors

I am quite confident that, if we were really determined, we could find that missing 0.57 acres—or most of it, at least. Every time I stand under the blazing in some park or other, watching my sons play football or soccer, I think to myself: this place could really use some more trees. I really am not sure the 0.57 acres/person goal is that out of reach.[18] Nevertheless, there are also some reasons for pessimism—certain factors that push in the other direction. Some of these are relatively minor. For example, trees are clearly the best carbon sequestration devices, but all plants sequester carbon. My live oak stores 756 pounds of carbon, according to my calculations. But, if the oak were not there, and the area under its canopy were covered in, say, grass, this,

---

[17] Admittedly, once the new forests reach maturity, and the soil becomes saturated with carbon, we will have to adjust our calculations of carbon sequestration per acre downward, by a factor of roughly 20%. But that would not become a factor for a number of decades.

[18] For a useful assessment of the possibilities of carbon sequestration by urban trees, see D. Nowak and D. Crane, "Carbon Storage and Sequestration by Urban Trees in the USA," *Environmental Pollution* 116, no. 3 (2002): 381–89.

too, would store a certain amount of carbon. Even so, this would be minuscule compared to the carbon stored in the tree. To see this, all we need to do is mentally compare the weight of the tree with the weight of the grass that might have grown in its place if the tree were not there. This picture does not change substantially if we replace the grass with crops such as wheat, corn, barley, or maize. Sometimes one sees grasslands that are used for the grazing of cattle being touted as part of the solution to climate change.[19] No matter how you spin it, in terms of carbon-sequestering capacities, grasslands are a dreadfully poor substitute for forests. The more we make grasslands like forests, by adding stands of trees to them, for example, the better they become. But that just makes my point. From the point of view of climate mitigation, forests are best and other types of land are better the more they resemble forests.

Second, there is a more substantial worry concerning the albedo effect of forests. This is a reflectance issue. In Chapter 4, we talked about the reflectance capacity of the Artic, Antarctic, and Greenland ice caps: being white, they reflect heat back out into space, and if they were to be replaced by darker oceans waters, more of this heat would be absorbed, thus pushing up global temperatures further. Ice caps have a high albedo, and that of oceans is much lower. A similar issue arises through afforestation. In northern climes, for example, tree leaves (i.e., needles) tend to be quite dark, and thus will absorb heat. Imagine a pristine snowscape, a pure white expanse. Now imagine this carpeted with dark spruces or pines. The snowscape would reflect heat. The dark leaves will absorb it. This is an albedo effect—something that counts against the ability of trees to mitigate global warming.

There is no suggestion that the albedo effects of afforestation will, in general, outweigh the carbon sequestration efforts of the newly forested lands. In a study of afforestation on New Zealand's North Island, for example, Kirschbaum et al. conclude that, "Albedo and carbon-storage effects were of similar magnitude for the first four to five years after tree planting, but as the stand grew older, the carbon storage effect increasingly dominated."[20] Nevertheless, one should expect some net reduction

---

[19] See, for example, Lela Nargi, "Can Cows Help Mitigate Climate Change? Yes They Can!" *JSTOR Daily*, https://daily.jstor.org/can-cows-help-mitigate-climate-change-yes-they-can/.

[20] M. Kirschbaum, D. Whitehead, S. Dean, P. Beets, J. Shepherd, and A. Ausseil, "Implications of Albedo Changes Following Afforestation on the Benefits of Forests as Carbon Sinks," *Biogeosciences* 8, no. 12 (2011): 856.

in the climate mitigation potential of afforestation. This does suggest we might want to be quite particular about where we focus our afforestation efforts. Albedo effects will be greatest at higher latitudes, where there is the possibility of snow being obscured by dark pine needles for significant portions of the year. In temperate and tropical regions, the albedo effects of afforestation will be much less significant. Mykleby et al. have traced a "boundary through North America where afforestation results in a positive equivalent carbon balance (cooling) to the south, and a negative equivalent carbon balance (warming) to the north."[21] The proposal defended in this chapter is to afforest land currently used for grazing and the growing of feed crops. A pristine snowscape is not the sort of terrain which normally provides a good candidate for either crop or grazing land. Therefore, while the proposed afforestation might involve some small lowering of albedo—if, for example, the leaves of trees are darker than the grass or crops—but not the kind of dramatic lowering involved in the transformation of a snowscape to forest. All afforestation defended in this chapter would occur on land that falls comfortably within Mykleby et al.'s positive equivalent carbon balance territory.

The third negative consideration turns on the role of trees in emitting methane—as we have seen, a much more powerful GHG than carbon dioxide. We know that some tress will emit methane under certain conditions. This is particularly true of trees growing in wetlands. But the study of this phenomenon is in its infancy. There is no suggestion that tree methane-emitting proclivities of trees outweigh their carbon-sequestering habits. In most cases, the consensus is that the latter dwarfs the former. However, we may have to make certain adjustments to the net GHG mitigation provided by forests. How much these adjustments are going to be, no one yet really knows. Further study needs to be done, and until it is, I shall leave this worry to hover furtively in the back of my mind.

There are virtually no scientists who think these countervailing considerations will be (anywhere near) enough to undermine the status of trees as carbon sinks. The overwhelming current consensus is that trees are, overall, carbon sinks—and are so to a degree that dwarfs the sequestering capacities of cropland, clearly outweighs any possible albedo effect, and almost

[21] P. Mykleby, P. Snyder, and T. Twine, "Quantifying the Trade-off between Carbon Sequestration and Albedo in Mid-latitude and High-latitude North American Forests," *Geophysical Research Letters* 44, no. 5 (2017): 2493–2501.

certainly outshines their role in emitting GHGs. This is why reforestation is such a big part of the attempts of many countries to meets their obligation under the Paris Agreement. The Intergovernmental Panel on Climate Change (IPCC)'s 2019 Special Report assumes we will need a "9.5 million km2 increase in forests by 2050 relative to 2010," and this is only in conjunction with some rather hopeful speculations about other energy pathways.[22] China has a stated a goal of planting new forests in a total area that is more than four times the size of the United Kingdom. The European Union is moving toward allowing countries to include forest planting in their plans to fight climate change. In the United States, California now allows forest owners to sell credits to CO_2-emitting companies, and other states are considering following suit.

Perhaps most obvious, and most important, is a fourth reason for caution. Not all the land we currently use for grazing will sustain trees. Parts of the Great Plains of the United States, for example, are too arid. The same may be true of parts of the South American Pampas plains. The proposal I am defending is, of course, not that we should try to grow trees where they will not grow, but try to grow them everywhere they will—and also be quite imaginative about this. Even on the Great Plains, *riparian* forests, more than a mile wide in some cases, used to grow all along the banks of streams and small waterways, like marbled veins of the heartland.[23] That's where we put trees back: not where they will not grow, but where they will. But this suitability factor does mean we will, to some extent, have to revise downward the amount of land made available for afforestation through elimination of the meat industry. As we shall see toward the end of this chapter, this probably doesn't matter at all.

There and Back Again

Far more worrying than albedo effects, offsetting methane emissions, and suitability of land is the shifting location of "there" I mentioned earlier. This can be a good thing, but it is more commonly turning out to be a bad thing.

---

[22] This, the IPCC advocates, is to be taken from current pastureland. IPCC, "Global Warming of 1.5°C," (SR15), C 2.5, https://www.ipcc.ch/site/assets/uploads/sites/2/2019/06/SR15_Full_Report_High_Res.pdf.

[23] See E. West and G. Ruark, "Historical Evidence of Riparian Forests in the Great Plains and How That Knowledge Can Aid with Restoration and Management," USDA Forest Service/National Agroforestry Center, September 2004.

It all depends on the direction of shift. "There" refers to the total carbon footprint of—the total amount of carbon dioxide produced by—an individual person. In the United States, according to the World Bank, that stands at 16.5 tons. Suppose that is my carbon footprint, and suppose, for the time being, I have no family to complicate the calculations. But, I would point out, I have a garden with trees that offset this to the tune of 0.5 metric tons. Given this is so, since I own the land and the trees, my actual carbon footprint is 16 tons. But suppose my neighbor sneaks in one night and cuts down all my trees. It wasn't my fault. I had nothing to do with it and, indeed, object most strenuously. It doesn't matter: my carbon footprint is back up to 16.5 tons. Let's now bring my family—who I've apparently just remembered I have—back into it. Our carbon footprint has gone up by 0.5 tons. This means that the carbon footprint of each of us has gone up by 0.125 tons. The moral is that every time a tree is cut down, your net carbon footprint goes up—even though you had nothing to do with it. In fact, the net carbon footprint of everyone goes up, albeit marginally.

In the battle to get climate change under control, trees are our best friends. And we are, it goes without saying, our own worst enemies. We are our own worst enemies because we are killing off our best friends at unprecedented rates. According to the World Bank, between 1990 and 2016, the world lost 502,000 square miles (1.3 million square kilometers) of forest.[24] Let's get back into calculating mode. One square mile is equal to 640 acres. Therefore:

$$502,000 \text{ square miles} = 321,280,000 \text{ acres}$$

So, if the World Bank is correct, we have lost over 321 million acres of forest in the 27-year period between 1990 and 2016. Earlier calculations employed the assumption—in the conservative range of EPA and USDA estimates—that an acre of afforested land can sequester roughly 5 metric tons of carbon dioxide per year. One of those tons consisted of carbon dioxide sequestered in the soil. Afforested land, however, is different from historically mature, historically forested land. In the latter, the soil will typically have reached the carbon saturation point and so can absorb no more. In calculating the opportunities for carbon storage lost, therefore,

[24] World Bank Blogs, https://blogs.worldbank.org/opendata/five-forest-figures-international-day-forests.

I shall count only the amount of carbon dioxide sequestered in the trees themselves—4 metric tons per acre per year. Given this assumption, the size of the lost opportunity for carbon sequestration can be found by multiplying the number of acres lost by the amount of carbon dioxide sequestration per acre per year:

$$321 \text{ million} \times 4 = 1{,}284{,}000{,}000$$

That is, 1 billion, 284 million metric tons—or nearly 1.3 gigatons—of carbon dioxide could have been sequestered by these trees every year had they not been cut down. Every year from now on, we lose the ability to sequester nearly 1.3 gigatons of carbon dioxide because we cut down these trees. Nor is this deforestation abating. On the contrary, it is, if anything, getting worse. If that is not bad enough, the amount of carbon released in the atmosphere from this policy of slashing and burning our way through the world's forests is truly terrifying.

To get a rough impression of this figure, let's work again with 135 trees per acre, containing, as in the case of my live oak, 756 pounds of carbon each. Thus, the total carbon contained in the acre of forest will be:

$$135 \times 756 \text{ pounds} = 102{,}060 \text{ pounds}$$

A metric ton is equal to 2,204 pounds. Therefore, to convert this to metric tons, we divided 102,060 by 2,204:

$$= 46.3 \text{ metric tons}$$

So we can assume that an acre of forest will have a carbon weight of 46.3 tons. Let's just say 46. To find the total carbon weight of the trees we have cut down between 1990 and 2016, we simply multiply 46 by the total number of acres deforested:

$$46 \times 321 \text{ million} = 14{,}766{,}000{,}000$$

That is, a little under 15 billion tons of carbon will have been contained in the land we have deforested between 1990 and 2016. This carbon will return to the atmosphere, where each atom of carbon (of atomic weight 12) will combine with two oxygen atoms (combined atomic weight 32), to yield a weight of carbon dioxide that is:

$$14.7 \text{ billion} \times 3.6 = 52.9 \text{ billion metric tons}$$

Let's say 52 billion. If this is correct, the trees we cut down during between the years 1990 and 2016 have added 52 billion tons—52 gigatons—of carbon dioxide to the atmosphere. That is equivalent to a little over 3 gigatons of carbon dioxide per year. This assumes that all of the trees destroyed are burned or left to rot. Of course, this assumption is, in general, false. But, depending on place and circumstance, it is often not far from the truth: it is typical of deforestation in many parts of the world. And even when the timber is used for other purposes—construction, for example—it is still true that many parts of the tree are burned or left to rot. Nevertheless, we will have to adjust these estimates depending on the percentage of timber used for construction or other purposes.

Enough, however, of my amateurish calculations. The acid test is, of course, whether they stack up against those who do this sort of thing for a living. In comparison, my figures are actually on the conservative side. The World Resources Institute puts it nicely: "If deforestation were a country, it would rank third in carbon dioxide equivalent emissions, only behind China and the United States of America."[25] The World Resources Institute (WRI) estimate differs slightly from mine: they argue that deforestation is responsible for 4.8 gigatons of CO_2-equivalent emissions per year. This is higher than my 3 gigaton estimate. The difference, however, is not surprising, for two reasons. First, the WRI figures are calculated for the years 2015–2017, whereas I have simply averaged out across the period of 1990–2016. Deforestation, however, has been increasing steadily over this period. In 2001, for example, it was estimated to add to the atmosphere just over 2 gigatons of CO_2-equivalent gases, in 2016 that figure had jumped to just below 6 gigatons. Secondly, I have made some very conservative assumptions, particularly with regard to carbon weight per acre. My estimates are, I suspect, almost certainly a best-case scenario, and matters are likely to be quite a bit worse than my calculations reflect. Thus, when assessed over the same time period—say

---

[25] World Resources Institute, https://www.wri.org/blog/2018/10/numbers-value-tropical-forests-climate-change-equation.

2015–2017—and when adjusting for different operating assumptions, my calculations and those of the WRI are relatively consistent.

While we are on the subject of best-case scenarios, it should be noted that other estimates of global deforestation are higher than that of the World Bank. Some are much higher. According to a satellite study conducted by the University of Maryland, global deforestation accounted for an astonishing 888,000 square miles (2.3 million square kilometers) between 2000 and 2012.[26] If this is correct, we can increase the figures by a factor of 1.76 (and that over a shorter time frame). My figure of 52 billion metric tons of carbon dioxide released into the atmosphere through deforestation transforms into a figure of 91.5 billion metric tons. The annual figure jumps to 5.28 gigatons of carbon released per year. My figure of 1.3 gigatons of lost sequestering ability per year jumps to nearly 2.3 gigatons of annual lost sequestering ability. Whether the lower- or higher-end estimates turn out to be true is, in a sense, immaterial. Either range of estimates is grotesque, and potentially civilization threatening, if we continue with our slashing and burning business-as-usual. There is no easy solution to this problem because of the nature of the atmosphere.

The Tragedy of the Atmospheric Commons

The global figures make it clear that all the reforestation we can muster in the United States is not going to have much effect if, say, Brazil keeps doing what it seems to be doing. Or if Malaysia and Indonesia, understandably seduced by the global demand for palm oil, continue doing what they are doing. Every time an acre of forest is cleared, the average net carbon footprint of everyone in the world goes up, even though most have no inkling of what was happening, and haven't changed their behavior one bit. At the heart of this seeming paradox is the idea of the *commons*. Forests may belong to nations or individuals, but the atmosphere belongs to everyone—and therefore to no one nation or person in particular. The almost inevitable result of this combination of ownership and absence of ownership is *tragedy*.

---

[26] NASA, Landsat Science, https://landsat.gsfc.nasa.gov/when-trees-fall-landsat-maps-them/. It is true that, as the study also notes, in the same time period new forests grew by 309,000 square miles, which will provide some welcome offsetting in the form of sequestration. My focus here, however, is simply on the amount of GHGs produced by the deforestation itself.

The term "commons" was employed in this form, and combined with the idea of tragedy, in a famous paper by Garrett Hardin.[27] A *common* was an area of land in or near a medieval village where everyone grazed their livestock. The land belonged to everyone, and therefore no one. Suppose you are a medieval subsistence farmer who has just inherited a few more cattle. The common has, admittedly, been a little busy lately, and the grass is wearing a little thin. But you have a right, as does everyone, to graze your cattle on it, and if you this will benefit you—either through sales of the cattle or through sales of cattle-related products such as meat, milk, and cheese. So you are better off if you graze your cattle. To put the matter in the standard terms of economics, grazing your additional cattle has a *positive utility* for you, which is equal to the proceeds from the sale of the animal or its products. Let's give this positive utility the value of 1 on some arbitrary scale of utility. It also has a *negative utility* for everyone, including you, that consists in the overgrazing caused by your additional cattle. However, this negative utility is a problem for everyone. And, so, the negative utility is divided up equally between everyone. If the negative utility were 1, for example, then the negative utility for you would be 1 divided by all the people on whom it has a negative impact. The negative utility does not have to be 1, but as long as the negative utility divided by the number of people impacted by it is less than 1, the positive utility of the additional grazing for you will outweigh the negative utility for you. The larger the number of people, the more likely it is that the negative utility for you is less than 1. Therefore, it is, overall, very likely that the positive utility for you will outweigh the negative utility for you. The same is true for anyone else who happens to have acquired a few more head of cattle. Therefore, there will be a strong pressure toward overgrazing. This is the logic of the commons. The result, if Hardin is right, is a kind of *tragedy* in the true sense of the word. Tragedy, on most classical theories, is more than merely misfortune. It involves a sense of inevitability. The tragic hero, or the audience, recognizes that the approaching doom is inevitable and unavoidable, that the outcome will be the same no matter what actions he or she takes.

The commons, in Hardin's sense, is any shared resource. This includes the atmosphere and also the climate. Cutting down vast areas of forest in order to facilitate development of a cattle ranch may undoubtedly make sense to— that is have positive utility for—the owner of the land who becomes rich as a result of using it in this way. It may make sense to a nation within whose

---

[27] Garrett Hardin, "The Tragedy of the Commons," *Science*, December 13, 1968, 1243–1248.

borders the forest lies, especially if we do not look too far into the future. It may increase the nation's gross national product, may help balance its trade deficit, and may provide jobs and incomes, and tax revenues, and all sorts of other things that nations are very fond of having. Deforestation, for the nation, can have positive utility in this sense. It also has negative utility of various sorts, including its impact on an atmosphere already choking on carbon dioxide and other GHGs. However, this negative utility is negative utility for the world, and not just the individual or the nation. The result will be a strong pressure—on the individual or the nation—toward deforestation.

The only way to address this beckoning tragedy is to understand that a nation or individual in possession of large tracts of forest is a nation or individual in possession of a valuable resource in the fight against climate change, and that they should be compensated properly for the role they are playing in this fight. If a certain area of tropical forest in Indonesia, for example, has a certain value because of the palm oil it can produce, then the owner of that area of forest—whether a nation or an individual—will have to be compensated by that amount for maintaining the forest rather than cutting it down. If a certain tract of the Amazon rainforest is deemed to have a certain market value because of the livestock it could support, the owner of that forest, whether individual or nation, should be compensated in this amount for maintaining the forest. This kind of international cooperation can be achieved in various ways. An industry in the United States, for example, should be able to buy carbon credits by paying the owner of a tract of tropical forest to maintain their forest rather than cut it down—paying them the agreed value of the money they could have made if they had deforested. This kind of purchase of carbon credits by US industries is becoming increasingly common. Moving from the individual or corporate to the national, the preservation of forests should be at the forefront of trade agreements struck up between nations. Foreign aid can be targeted, and more than justified, when the recipient of the aid agrees to preserve their forests or regrow them to their past state. In doing this, we move away from an *aid* model of international cooperation to a *trade* model. In such circumstances, we are no longer talking about aid, but payment for a vital service rendered.

This is all, without doubt, naïve. It would require a level of international cooperation—not to mention a rational economy—undreamed of at present. The lead would have to be taken by the developed countries who would have to be willing to (1) tax carbon heavily enough in their countries to (2) make the price of the carbon credits they have to buy in order to offset their emissions (3) sufficiently attractive to the owners of forests to preserve them. This general idea is captured in Article 6 of the Paris Agreement. Article 6 is, admittedly, the most controversial part of the Agreement. But it is difficult to see how the tragedy of the atmospheric commons can be avoided in the absence of a so-called *carbon market*, where this kind of reward for possession and maintenance of the world's forests can be implemented.

The Big Picture

Precision of calculation has not, as I mentioned earlier, been a priority of this chapter. I suppose you might wonder why a philosopher has spent so much time doing mathematics in this chapter. I have to admit: in part, for the fun. It's been very simple math anyway. And, even so, in developing the arguments of this chapter, I haven't relied on any calculations of my own, preferring to calculate wherever I can, and then check the results with the professionals. I have, as I mentioned at the beginning of the chapter, been far more concerned with painting a picture, or the contours of a picture, and my calculations are merely rough sketches of shapes, lines, and edges—intended to give a broad idea of the potential of reforestation rather than a precise quantitative accounting of this potential.

Cutting through the details, there are really just two figures that are crucial to this picture. *First*, how much carbon dioxide will an acre of afforested trees sequester per year? The EPA estimates between 2.2 and 9.5 metric tons. The USDA tells us it is between 2.7 and 7.7 metric tons. In the picture I painted, I worked with a figure of 5 metric tons—4 metric tons sequestered by the trees, and 1 by the soil—that is toward the conservative end of the EPA estimate and almost in the middle of the USDA estimate. But you can, if you are so inclined, pick any value between 2.2 and 9.5 and have at it yourself!

Second, how many acres are available for afforestation? This is where it gets interesting. In a recent study, Bastin and colleagues employed data from 78,744 photo-interpretation measurements of tree cover across the world. They then used a machine-learning algorithm to ascertain which parts of the plant's surface could support afforestation. Through examination of how climatic, edaphic (i.e., relating to soil), and topographic variables drive variation in tree cover, they estimated the number of additional acres capable of supporting forest under current climatic conditions. Their conclusions are striking: outside of existing forest, urban centers, and agricultural land, they estimate that the earth could support another 2.2 billion acres (0.9 billion hectares) of afforested land.[28] If we assume that an acre of afforested land will sequester 5 metric tons of carbon dioxide a year, then this will amount to 11 metric gigatons of carbon dioxide stored by this newly afforested land per year.[29] This is not far off from one-third of the total annual amount of anthropogenic carbon dioxide emissions in 2019.[30]

Even more striking, Bastin's figures are intended to exclude not only existing forest and urban centers but also existing agricultural land.[31] Imagine, therefore, if we were also able to turn over a significant portion of existing agricultural land—land used for animal grazing and the growing of animal feed crops—to this afforestation project. This land will often—not always, but often—be prime real estate for afforestation: if it can sustain crops, it can also, probably, sustain trees. Some grazing land may not be suitable for afforestation, but much will. Mercifully, I don't intend to do any more calculations here—my brain's math budget is spent. But our earlier examination of just how much land in the United States is currently dedicated to animal agriculture—whether through grazing or the growing of feed crops—provides us with a striking indication of the global potential of this strategy.

As is always the case with trees, their potential is a double-edged sword. Bastin and colleagues also estimate that on our current climate trajectory, "the global potential canopy cover may shrink by ~223 million hectares by 2050, with the vast majority of losses occurring in the tropics." This is carbon sequestration potential that we cannot afford to lose, and time is, clearly, of the essence. The trajectory needs to be broken—and broken soon. The most effective, and the most decisive, step we can take to mitigate climate change in the short term is to afforest wherever we can and whenever we can. But wherever and whenever are shifting locations. If we stop eating meat, the scope of wherever and whenever expands dramatically. With this expansion,

[28] J-F. Bastin, Y. Finegold, C. Garcia, D. Mollicone, M. Rezende, D. Routh, C. Zohner, and T. Crowther, "The Global Tree Restoration Potential," *Science* 365, no. 3448 (2019): 76–79.

[29] Bastin estimates the sequestering of 205 gigatons of carbon (note: not carbon dioxide) by the time this afforested area reaches maturity.

[30] Around 36.8 gigatons. See P. Friedlingstein, M. Jones, M. O'Sullivan, R. Andrew, et al., "Global Carbon Budget 2019," *Earth System Science Data* 11 (2019): 1783–1838. See also G. Peters, R. Andrew, J. Canadell, et al., "Carbon Dioxide Emissions Continue to Grow amidst Slowly Emerging Climate Policies," *Nature Climate Change* 10 (2020): 3–6, https://doi.org/10.1038/s41558-019-0659-6;.

[31] I say "intended." As we shall see in Chapter 12, there is a dispute over how successful Bastin and colleagues were in implementing this exclusion. As we shall see, this, if anything, strengthens my case.

the value of afforestation in the mitigation of climate change can transform from highly significant to something potentially game-changing.

Therefore, I put it to you that probably the best thing we can do for the planet, to save it from the impending climate threat, is to (1) *stop eating meat* and (2) *afforest land* wherever and whenever we can—including, crucially, land formerly allocated to the raising of livestock, whether grazing land or cropland dedicated to livestock feed. Of course, it goes without saying that we should target carbon-dioxide equivalent emissions in all their forms and from all of their sources, from transport, industry, and domestic sectors. But abandoning meat and afforesting the land would, on its own, go a *long way* toward erasing our carbon footprint, and give us the breathing space we need to get improve our renewable technologies to the point of viability, which would, in turn, allow us to properly address these other emission sectors. In the short to medium term, these two modest steps provide the simplest and most decisive way of mitigating climate change.

Playing a significant—possibly game-changing—role in arresting climate change is, however, only one of the benefits of these twin policies of abstention from meat and afforestation of land. In the next four chapters, I shall argue that they can play a similarly vital, and perhaps decisive, role in the two other major environmental threats that we face today: *mass extinction* and *pestilence*. It is to the issue of extinction that we now turn.

8

The Great Dying

Concepts and Crimes

The expressions *great extinction* and *mass extinction* are often used inter-changeably. For my purposes, however, it is useful to draw a distinction be-tween two ways of thinking about, and measuring, extinction. Thus, by *great* extinction, I shall mean an event in which a given percentage of planetary species become extinct. A *mass* extinction, however, occurs when the actual *rate* of extinction exceeds the normal background rate by a certain speci-fied margin. Adding specific numbers to these definitions would be partly stipulative and therefore partly arbitrary, and it is not, in fact, that important for my purposes. What is important is that there are two distinct concepts we might employ in talking about extinctions. One—the concept of a great extinction—measures extinction in terms of *numbers*, or *percentage*, of spe-cies that die out. The other—the concept of a mass extinction—measures ex-tinction in terms of the *rate* at which species die out. All great extinctions will be preceded by mass extinctions, but not all mass extinctions necessarily lead to great extinctions.

Perhaps, however, you are someone who likes specific numerical values—and admittedly, these would be useful in pinning down the kinds of phe-nomena we are dealing with. It is possible, I think, to provide some reasonably principled numbers. Consider, first, the idea of a great extinction. There have been five unquestionably great extinctions on earth: the *end-Ordovician*, the *late-Devonian*, the *end-Permian*, *end-Triassic*, and the *end-Cretaceous* extinctions. The end-Ordovician, or Ordovician-Silurian, extinction saw the demise of approximately 85% of the earth's species. The late Devonian ex-tinction was slightly less severe, accounting for roughly 75%–80% of all spe-cies. The end-Triassic, or Triassic-Jurassic, extinction event was of similar severity, ending roughly 75% of all species. The end-Cretaceous extinction, or K-T extinction, the most famous of great extinctions as it saw the end to the tenure of the dinosaurs, displayed similar levels of awfulness, accounting for approximately 80% of all species. The greatest—that is, worst—of them

World on Fire. Mark Rowlands, Oxford University Press. © Oxford University Press 2021.
DOI: 10.1093/oso/9780197541890.003.0008

all, however, was the end-Permian, or Permian-Triassic extinction, which nearly saw the end of all life on earth. Over 90% of all species became extinct, rising to over 95% in marine species. Nearly every single species of tree disappeared, and never returned. The end-Permian extinction event is sometimes known as *the great dying*.

There is, today, much discussion of whether we are in, or entering into, a sixth great extinction: the *Holocene* extinction—or, perhaps, the *end-Holocene* extinction if our current age is indeed going to be renamed the *Anthropocene*.[1] If the mooted extinction is to bear the label "sixth," then it is not unreasonable to suppose that it would have to be of the same kind of level of severity as the big five extinctions listed earlier. This suggests we might set the threshold for a great extinction at somewhere in the region of 75% of species lost. Much less than this might raise the question, in some minds at least, of whether the epithet "great" is really appropriate.

Any numerical criteria for qualifying as a mass extinction would, perhaps, be slightly less principled, but still not unreasonable. Recent estimates of the current rate of extinctions typically range from several hundred to several thousand times the normal background rate. It is difficult to find anyone who thinks current extinction rates are less than 100× the background rate. Ceballos et al. have argued that *even on the most conservative assumptions imaginable* the current rate of mammalian extinctions is at least 100× the normal background rate.[2] Other classes—notably amphibians—have fared particularly badly in recent decades, and for them estimates of extinction rates are much higher—as high as 45,000× the normal background rate.[3] It we wanted to assign a specific number as a threshold, we might plump for 1000×. This figure is not uncommon, and it occupies a fairly reasonable mid-range point in the spectrum of possibilities, although, other than that, it is

[1] For a gripping popular account, see Elizabeth Kolbert, *The Sixth Extinction: An Unnatural History* (New York: Henry Holt, 2014). Also equally gripping is Edward O. Wilson, *Half Earth: Our Planet's Fight for Life* (New York: Liveright, 2016). Very worthy of mention, and written almost 20 years before these, there is Richard Leakey and Roger Lewin, *The Sixth Extinction: Patterns of Life and the Future of Humankind* (New York: Doubleday, 1995).

[2] G. Ceballos, P. Ehrlich, A. Barnosky, A. Garcia, R. Pringle, and Y. Palmer, "Accelerated Modern Human–Induced Species Losses: Entering the Sixth Mass Extinction," *Science Advances* 1, no. 5 (2015): e1400253, doi:10.1126/sciadv.1400253. The conservative assumptions in question are (1) that the background rate of mammalian extinctions is 2 extinctions per 10,000 species per 100 years, which is twice as high as widely used previous estimates coupled with (2) exceptionally stringent criteria for when a species can be regarded as extinct.

[3] See, for example, M. McCallum, "Amphibian Decline or Extinction? Current Declines Dwarf Background Extinction Rate," *Journal of Herpetology* 41, no. 3 (2007): 483–491.

not clear that this figure has anything, particularly, to recommend it over, say 100 or 10,000.

I can afford to regard these issues with something approaching equanimity. I shall be concerned with extinction in this chapter and the next, but part of the message I shall seek to convey is that extinction is not the be all and end all of the discussion of species and their decline. Rather, I shall argue that our concern with extinction can plausibly be regarded as a symptom of a deeper and more fundamental concern. At the core of this concern, I shall argue, is a *crime* of certain sort—one that we humans have committed and continue to commit.

Suppose, to get the dialectical ball rolling, there is something to the talk of a *sixth extinction*. If this is so, then unlike the other great extinctions, for which we obviously can't be blamed, we humans are looking distinctly guilty-as-charged for this latest one. There is a bloody dagger held, figuratively but tightly, in our sweaty, acquisitive little hands. Nevertheless, questions still arise. One in particular: *guilty of what*? I think it would be difficult to convict us of the crime of (bringing about a) *great extinction*. It is clear that the current state of nature today is nothing like the way it was during the five great extinctions. At least not yet. But the thing about extinctions is that they take time. The greatest of debacles rarely happen overnight. The word "event" is often coupled with the word "extinction," yielding *extinction event*. But this is misleading. If extinctions are events at all, they are so only on geological timescales. And geological timescales are, in comparison with our human temporal frames of reference, rather vast. The late Devonian extinction event, for example, was really a series of extinction pulses, spread out over 20–25 million years. We, anatomically modern, humans have been here for 200,000 years, tops. So it would be unreasonable to expect us to have orchestrated a great extinction in the short time available to us. We might, possibly, get done for *conspiracy*, although even that would be pushing it because conspiracy presupposes intention, and I suspect carelessness and selfishness lie at the heart of our failings rather than deliberate malice. But we certainly can't be found guilty of the crime of great extinction because there is, as yet, no great extinction.[4]

---

[4] How much time it will take for us to break into the category of great extinction is an interesting question, and it may happen a lot sooner than we think. A survey conducted in 1998 by the American Natural History Museum indicated that 70% of biologists believe that as many as of 20% of all species alive (then) will be extinct within the next 30 years. A third of biologists think that as many as half the species on earth will die out in the next 30 years. That is a mere 8 years from now. Unless I get a bit of a

The charge of mass extinction, however, has an altogether more plausible sheen to it. Of that we may well be, as they say, *bang to rights!* Nevertheless, given the partly stipulative nature of any numerical threshold for mass extinction, I am sure there is a lot of wiggle room here that a competent defense attorney for the human race might be able to exploit. My role in this chapter, however, is one of prosecutor. And as any good prosecutor knows, the most important thing is to get the bastard you are trying found guilty—of something! Precisely what is something you can worry about later. Thus, I shall, first, begin with a little history, with a view to establishing a *pattern of behavior.*

A Bestiary of Barely Imagined Beasts

With mass extinctions, as with great extinctions, it is important to have a grasp of the appropriate timescale. Like great extinctions, mass extinctions are also prolonged and come in pulses. It is common to hear claim that we are, today, *entering* into a period of mass extinction. In fact, it is far more likely that, if there is a mass extinction event occurring today, we are in the *middle* of it rather than entering it. I say "middle" in a deliberately vague sense. What I wish to convey by this is not that there is some precise arithmetical middle of this mass extinction event that I have somehow managed to identify, and that is where we are. Rather, I mean to convey only that if there is a mass extinction event currently unfolding before our eyes, we are certainly not at the beginning of it and just as certainly not at the end.

Once upon a time, there were lions in America, from sea to shining sea. The American lion lived in North America during the Pleistocene period, from around 340,000 up until 11,000 years ago. It must have made a magnificent sight: almost half a metric ton of highly evolved, heavily muscled killing machine—25% bigger than the modern African lion—striding the grasslands, forests, and high mountain passes, all the way from Mexico to Alaska. Like many who made their way to these shores, the American lion was an immigrant. Europe, too, had its lions in those days, as did most of Asia. Genetic analysis has shown the American lion to be closely related to the European (or Eurasian) cave lion. Ranging, in its heyday, from the straits

---

move on with the writing this book, by the time it comes out we could be in great extinction territory. See http://www.mysterium.com/amnh.html.

of Gibraltar all the way to the Bering Sea, the European cave lion had first appeared in Eurasia around 600,000 years ago and remained there until it became extinct in roughly 13,000 BC. Not quite as large as the American lion, the European cave lion was still more than 10% bigger than the modern African lion. The American lion evolved out of its Eurasian cousin, with the latter having migrated across the Bering land bridge. When the bridge disappeared, due to rising sea levels, the lion became genetically isolated from its Eurasian cousin and evolved into a distinct, larger sister lineage.

The American lion was just one of many fabulous beasts on the North American continent at this time. Among its competitors was the *Smilodon*. Often going by its alias, the *sabre-tooth cat* or *tiger* (misleading because they were not closely related to tigers), the Smilodon was one of the larger members of the genus of sabre-tooth cats. Not the largest of this family—it had a larger South American cousin—it was nevertheless formidable. Smilodon lived in the Americas from roughly 2.5 million to 9500 years ago. Although it was around a foot shorter than the modern African lion, it was twice as heavy and built somewhat differently. It had a shorter torso, and a bobtail, in some ways reminiscent of today's lynx. However, it had a massive head and shoulders and, unlike the lynx, its spine sloped slightly downward to its hips.

Another resident was the *dire wolf* which lived in the Americas from roughly 125,000 to 9000 years ago. Somewhat bigger—by 10%–15%—than the largest of present-day wolves, remains of the dire wolf have been found across a broad range of habitats, including forested mountain areas, the savannahs of Southern America, and the grasslands of North America. If we go by the numbers of remains discovered, it is likely the dire wolf preferred the open lowlands that housed its typical prey—various species of large herbivore—and for these there is some evidence that it would compete with both the Smilodon and the American lion.[5]

The *short-faced bear* was one of the largest land-based carnivores ever recorded, with the heaviest specimens weighing in at over 950 kilograms—not far short of a metric ton. When walking on all fours, the shoulders of this animals stood 6 feet off the ground, and its height would more than double when it stood on its hind legs. Indeed, claw marks from the bear on cave walls have

---

[5] J. Coltrain, J. Harris, T. Cerling, J. Ehleringer, M-D. Dearing, J. Ward, and J. Allen, "Rancho La Brea Stable Isotope Biogeochemistry and Its Implications for the Palaeoecology of Late Pleistocene, Coastal Southern California," *Palaeogeography, Palaeoclimatology, Palaeoecology* 205, no. 3–4 (2004): 199–219.

been discovered 15 feet off the ground. It was able to run at speeds of up to 40 miles per hour and, among other things, may have hunted mammoths: what seem to be teeth marks from the short-faced bear have been found on mammoth skeletons.[6] It was probably descended from the Eurasian cave bear and, like the American lion, evolved into a distinct, larger, species following its journey across the Bering land bridge and subsequent genetic isolation. It lived in North America from around 1.8 million until 11,000 years ago.

The large species of herbivore on which these formidable carnivores preyed were equally astonishing. Several species of giant ground sloth, *Megatherium*, larger than an elephant, roamed the land in those days, before becoming extinct around 10,000 years ago. There was also the *Pleistocene bison*, the ancestor of today's bison, but as much as 25% larger. It, too, became extinct around 10,000 years ago. There was the *Camelops*, a species of huge, extinct camel that populated North America during the Pleistocene, becoming extinct around 11,700 years ago. There were also various species of prehistoric horse and prehistoric deer. Until 10,000–11,000 years ago, *American mastodons* roamed the plains. Most famously, there were *mammoths*—woolly mammoths in the far north and Columbian mammoths ranging from the Canadian border as far as central Mexico. Weighing up to 10 metric tons (22,000 pounds), these were much larger than the modern African elephant, which generally tops out at around 13,000 pounds. The mammoth became extinct in North America around 10,000 years ago.

An astonishing bestiary, barely imaginable to us, perhaps. But try to imagine what it must have been like to share the land with these wondrous creatures. In the first of the *Jurassic Park* films—Park, not World, you know the old ones with Sam Neill instead of Chris Pratt—there is a moment, early on in the film, when the characters' Jeep comes to the top of a rise, and they see thousands of dinosaurs spread out as far as the eye can see. The reaction that Spielberg is going for is clear: *that would be so cool!* I confess, I think I would have a similar reaction to alighting at the top of a rise and seeing the beasts of the Pleistocene arrayed like this before my eyes—as long as they stayed at a safe distance. Because of their loss, the world of today is, I think, massively impoverished: a smaller, sadder, markedly less *cool* world. And while we can't be blamed for the demise of the dinosaurs, that of these Pleistocene creatures

---

[6] D. Gillette and D. Madsen, "The Short-Faced Bear *Arctodus simus* from the Late Quaternary in the Wasatch Mountains of Central Utah," *Journal of Vertebrate Paleontology* 21, no. 1 (1992): 107.

was almost certainly our doing. We killed them all. This, to me, seems to fall into the category of *some sort of crime.*

There Goes the Neighborhood

Did we really do it? Are they really gone because of us? Are we really guilty of these historic crimes of *speciocide*—of crime after crime of ancient species extermination? No one doubts that what is known as the *Quaternary extinction event* is a thing. Like most extinction events, this "event" is really many events, a series of pulses of extinctions that occurred during the late Pleistocene period. The pulses began in earnest around 130,000 years ago—at the beginning of the late Pleistocene—and reached their apogee between 13,000 and 8000 years ago. All the animals listed in the previous section died during this latter period, the transition between the Pleistocene and Holocene ages.

While no one disputes the existence of this Quaternary extinction event, the reasons for it are still disputed. There are two main hypotheses. One is the claim I have advanced. We killed them all. This is known as the *hominin paleobiogeography hypothesis.* The other is our old and most faithful exculpation. It was the climate change what done it, guv! This is the *climate change hypothesis.*[7] The climate change hypothesis is not unreasonable: the world, after all, only fully emerged from the last glacial period around 11,700 years ago. One thing we should definitely guard against is fallacy of *false dichotomy.* There is no reason to suppose that *either* the climate did it *or* Paleolithic humans were responsible. A more sensible approach might be to make the default assumption that both parties might have had a role to play, and then try to identify the evidence that would show the relative impact of each. It might be, for example, that a given species is weakened by one factor and finished off by the other. Remember also that the Quaternary extinction event spans more than 100,000 years, comprising a series of extinction pulses, and the primary mover in one pulse is not, necessarily, the primary mover in

---

[7] There were two other hypotheses that never really achieved similar sustained popularity. The first was *widespread disease.* This, however, seems to require us to believe in the existence of a pathogen that afflicted a huge variety of genera, or afflicted prey common and essential to a huge variety of genera. This, it seems, would be an entirely unprecedented pathogen. The other was *meteor strike.* This hypothesis has two main drawbacks. First, there is a lack of clear supporting evidence. Second, even if true, it would collapse into the climate change hypothesis since the impact on the meteorite on the world have been through the climatic changes it induced.

another. Finally, if the question under consideration is whether we are guilty of the historic crime of speciocide, it is far from clear that climate change can absolve us of this guilt as long as we also played some nonnegligible role. If I break into my neighbor's house to rob him, and he dies of a heart attack from the shock, I cannot absolve myself of responsibility by pointing out that his heart was a bit dodgy, and he really could have dropped at any time. Indeed, in some countries, I would be judged more harshly for this. This book is not the appropriate vehicle for trying to adjudicate the dispute between the climate change hypothesis and the hominin paleobiogeography hypothesis in general. In what follows, I shall merely outline the case for thinking that Paleolithic humans played a significant role in the late Quaternary extinction pulses, particularly the pulse of 13,000–8000 years ago that saw the end of the creatures described earlier.

The hominin paleobiogeography hypothesis achieved early popularity through the work of the geoscientist Paul S. Martin in the 1960s and 1970s.[8] Much more recently, Sandom and colleagues have conducted what is probably the most systematic analysis of extinction pulses that have occurred between c. 132,000 years ago and the late Holocene, c. 3000 years ago.[9] This analysis is based on comprehensive, country-by-country, continent-by-continent data on the distribution of all large mammal species (where a *large* mammal is defined as one weighing 10 kilograms or more) that became extinct during this period. The conclusion they reach is unequivocal: "The global pattern of late Quaternary megafauna extinction presents a clear picture that extinction is closely tied to the geography of human evolution and expansion and at most weakly to the severity of climate change."[10] In other words, fossil records of extinctions of various species of megafauna strongly track the arrival of humans in the ancestral homes of these animals, but only very weakly track changes in climate in these regions.

The strategy of Sandom and colleagues is an obvious and reasonable one. Step One: examine the fossil records of a given species of megafauna to work

---

[8] See, for example, P. Martin, "The Discovery of America," *Science* 179, no. 4077 (1973): 969–974.

[9] C. Sandom, S. Faurby, B. Sandel, and J. Svenning, "Global Late Quaternary Megafauna Extinctions Linked to Humans, Not Climate Change," *Proceedings of the Royal Society B* 281 (2014), 1–9, https://royalsocietypublishing.org/doi/pdf/10.1098/rspb.2013.3254. For the specific case of Australia, see Saltré et al., "Climate Change Not to Blame for Late Quaternary Megafauna Extinctions in Australia," *Nature Communications* 7, no. 7 (2016): 105–111, https://doi.org/10.1038/ncomms10511. For a useful survey of the debate, albeit a little dated now, see P. Koch and A. Barnosky, "Late Quaternary Extinctions: State of the Debate," *Annual Review of Ecology, Evolution and Systematics* 57 (2006): 215–250.

[10] Sandom et al., "Global Late Quaternary Megafauna Extinctions," 6.

out when it became extinct, either globally or locally. Step Two: ascertain if there were any notable climatic changes around this time. Step Three: ascertain if the time of the extinction coincided with, or was briefly preceded by, the arrival of the humans in the ancestral range of the extinct species. Step 4: compare Steps 2 and 3. What Sandom and colleagues found, in effect, was that local extinctions of megafauna coincided far more closely with the arrival of humans than it did with the existence of climatic change.

For example, in North America, between approximately 11,500 and 10,000 years ago, the fossil record indicates that the area roughly corresponding to contemporary California rapidly lost 21 species of megafauna, including the giant ground sloth, the dwarf pronghorn, and the dire wolf. These extinctions followed closely on the heels of newly arriving humans. There was no noticeable climatic change at this time, and the vegetation that sustained herbivores was unchanged. In many cases, megafaunal herbivores went extinct, even though their preferred local diet remained plentiful. The same kind of pattern—although even more dramatic—is observed in South America, where high extinction rates, despite a relatively stable climate, were most striking. After the arrival of humans in South America, roughly 15,000 years ago, extinction rates were among the highest recorded in the late Pleistocene. There was some minor climatic fluctuation at this time, between humid and dry periods, and occasional shifts in the availability of savannah and forest habitats. However, dietary analyses of large South American herbivores such as the giant ground sloth, the mastodon, and the *toxodon*, a large hoofed-creature weighing over a ton—it looks a little like a horse had a baby with a hamster—indicate that these extinct species were flexible eaters, quite capable of altering their diet between grazing and mixed feeding depending on their immediate habitat. Moreover, many areas were maintained in stable condition, either as grassland or forest, throughout the whole period. The pampas, for example, existed as grassland through this entire period. But this did not stop many megafaunal species of these habitats from disappearing.

As Sandom and colleagues argued, whenever climatic variations are cited as the reason for species extinctions in the late Pleistocene, a useful point of reference is always Africa. For reasons we shall turn to shortly, Africa, during this time frame, witnessed no mass extinctions of the sort that plagued the rest of the world. There were, however, some climatic fluctuations during this period. Therefore, in assessing the claim that climatic changes caused extinctions in the Americas, Europe, or Asia, a good question to ask is whether these changes were greater than the ones experienced by Africa. If the African climatic changes did not cause extinctions, and the changes in other parts of the world were no greater, or less than, those in Africa, this casts serious doubt on the claim that climate change was primarily responsible for these other extinctions. In the case of the Americas and Australia, Sandom and colleagues argue, climate fluctuations were definitely no greater than those in Africa. In Eurasia the climate changes were somewhat more significant than those occurring in Africa, leaving open the possibility that climate change may have played some role here.

This comparative approach raises an obvious question: why were there no mass extinctions in Africa? After all, humans had been in Africa longer than anywhere else. Why wouldn't they have wiped out vast numbers of species, given all the time they had to do so? The answer has to do with the fact that humans originated in Africa and, thus, humans and the animals they hunted had millions of years to coevolve with each other. Essentially, the animals had developed behaviors required to cope with humans and their gradually emerging Paleolithic technologies. Contrast this with other parts of the world, where humans arrived—all *tooled up*, as they say back in London, with the best killing technology the Paleolithic age had to offer—and quickly imposed themselves upon animals who were completely unprepared for this. There are lions in Africa but none in Europe or America, because the lions of Africa, and their typical prey, had—on an evolutionary time scale—time to learn how to cope with the developing hominins that became humans. The lions of Europe and the Americas, and their typical prey, were afforded no such luxury. The newly arrived apes, sly and rapacious, wiped them from the face of the earth.

This same general pattern, Sandom and colleagues argue—extinction coinciding with the arrival of humans in the absence of substantial climate change—was repeated globally: "The pattern of extinctions closely followed, with increasing severity of extinction with reduced period of hominin-megafauna coevolution, notably with uniformly high extinction in areas where *H. sapiens* was the first hominin to arrive."[11] Only in Eurasia might climate have played a nonnegligible role, although even there, they argue, the role it played in extinction was not as great as that played by newly arrived humans. The picture is a depressing, if all too familiar, one: humans arrive

[11] Sandom et al., "Global Late Quaternary Megafauna Extinctions," 6.

and . . . there goes the neighborhood. In the demise of these astonishing animals, our guilt is likely: the beyond-reasonable-doubt kind of likely.

Not every mammoth died out 10,000 years ago. Most did. But a small band of mammoths survived until around 1650 BC—less than 4000 years ago. The Egyptian empire was already halfway through its three thousand-year existence by this point. The last stand of the mammoths took place on Wrangel Island, a 2900 square mile island off the northern coast of what is now Siberia. You might suspect, not unreasonably, that since it was an isolated group, the mammoths would have died out through inbreeding and lack of genetic diversity. However, the evidence for this is not at all clear cut.[12] Admittedly, their genetic diversity wasn't great, but the genetic diversity of mammoths, it turns out, was never that great anyway. And while it was a little lower on Wrangel Island, it is far from clear that it was the sort of diminution that would have led to rapid extinction. There were also no significant climatic changes at this time, certainly nothing in comparison to that which mammoths had endured during the millions of years of their existence. However, archaeological evidence indicates that humans first arrived on Wrangel Island around the same time that the mammoths there died out. While lack of genetic diversity might certainly have weakened the mammoths of Wrangel Island, it is likely that humans finished them off. In effect, we humans both started and finished the extinction of these majestic animals that had been on the planet for 5 million years. Wherever we go, things die. Sometimes they die forever. We humans are walking, talking extinctions events. The habitual extinguishing of species is one of our defining traits. *Homo sapiens* we call ourselves. If animals could name us, they would call us *Homo exterminatore*—man the destroyer. This is the *pattern of behavior* to which I alluded earlier.

There's Something Happening Here

These extinctions, one might think, were all rather a long time ago. Unfortunate, of course, but why not just let bygones be bygones? In fact, however, a paleobiologist from the distant future would be unlikely to

---

[12] V. Nystrom, L. Dalen, S. Vartantan, K. Liden, N. Ryman, and A. Angerbjorn, "Temporal Genetic Change in the Last Remaining Population of Woolly Mammoth," *Proceedings of the Royal Society B* 277 (2010): 2331–2337. https://doi.org/10.1098/rspb.2010.0301. Although, for an alternative, dissenting view, see E. Fry et al., "Functional Architecture of Deleterious Genetic Variants in the Genome of a Wrangel Island Mammoth," *Genome Biology and Evolution* 12, no. 3 (2020): 48–58, https://doi.org/10.1093/gbe/evz279.

distinguish what we are doing to the natural world today from the pulses of extinctions we enacted over the preceding tens of thousands of years. Today, most of us no longer pay much attention to the distinction between the Givetian, End-Frasnian, and Famennian extinctions—preferring to lump them together as "the" late Devonian extinction. The attitude of our future paleobiologist is likely to be similar. If there is a mass extinction event occurring today, we are not *entering* into it. We are *already* in it, and have been for quite some time. What we see today are simply segments of the same phenomenon. This is a pulsing wave of extinctions brought about, to a considerable extent, by humans: a wave whose nascent beginnings can be traced back 130,000 years. It is a wave to which climate change may certainly have given the odd diffident push, but lacking the concentrated, murderous impetus that was supplied by the bloody hand of human progress. It is a wave that surges on today.

The United Nations recently convened a body to look at the issue of extinction. This body is known as the Intergovernmental Science-Policy Platform on Biodiversity and Ecosystem Services (IPBES). It's a bit of a mouthful, admittedly. But, in a nutshell, IPBES is to extinction what the Intergovernmental Panel on Climate Change (IPCC) is to climate change. In May 2019, they released a summary of their assessment of the current state of the earth, with the full report following at the end of that year.[13] As you might have guessed, it makes salutary reading. Overall, they write: "Nature and its vital contributions to people, which together embody biodiversity and ecosystem functions and services, are deteriorating worldwide."[14] They note that 75% of the land surface has now been significantly altered by human activity. At the same time, 66% of the ocean area is now undergoing adverse cumulative impacts. And over 85% of wetlands have already been lost. The result is that "human actions threaten more species with global extinction now than ever before."[15] Specifically, the report estimates, around 25% of the animal and plant species assessed in the report are threatened. Extrapolating this percentage to all species suggests that "around 1 million species already face extinction, many within decades."[16] They estimate that there are currently around 8 million extant species. The figure of 1 million is, thus, a fairly

[13] "The Global Assessment Report on Diversity and Ecosystem Services," https://ipbes.net/global-assessment.

[14] "The Global Assessment Report on Diversity and Ecosystem Services," 10.

[15] "The Global Assessment Report on Diversity and Ecosystem Services," 11.

[16] "The Global Assessment Report on Diversity and Ecosystem Services," 11.

conservative extrapolation of the estimated 25% of threatened species in the sample assessed.

A mass extinction occurs when the actual rate of species extinction exceeds the normal background rate by a specified margin—earlier, I suggested 1000× the background rate as a reasonable threshold. To know whether this is happening, we first need to know the background rate of extinction. Calculation of this does not lend itself to any degree of exactitude. First you need to know the overall numbers of species in existence and, second, you need to know how long a species, on average, can be expected to survive. The background extinction rate is the number of existing species divided by average species lifespan. Lawton and May have estimated that, averaged out over all fossil groups, the average lifespan of species is somewhere between 500,000 and 5 million years.[17] In their calculations, IPBES assumed there are currently around 8 million species in existence. The IPBES figure of 8 million was based on a well-known paper by Mora and colleagues, which yields a mid-to-low-end estimation of 8.1 million.[18] However, other estimates vary considerably, ranging from a little over 2 million to more than 100 million. This vast disparity in estimates of species numbers reflects different methods and criteria of estimation. Most recent estimates, however, tend to coalesce in 5–20 million range. However, as IPBES acknowledges, no method of estimation is perfect, and this introduces a significant degree of uncertainty into our estimations of the normal background rate of extinction.[19] If we assume the kinds of species numbers estimated by Mora and colleagues, and also suppose the average life span of species is 1 million years, this means the background rate of extinction would be 8 species per year. If, on the other hand, average species lifespan was 2 million years, this would yield a background rate of 4 species per year. This sort of reasoning leads to the commonly cited background rate of extinction of 5–8 species per year. If we accept this figure, then a mass extinction would occur when species extinctions reached 5000–8000 a year.

The endemic uncertainty of species numbers and lifespans notwithstanding, it is clear that something very bad is occurring today, and it is very likely that this sort of threshold has been either breached or that we are coming very close to it. For example, based on a comparison of extinction

[17] J. Lawton and R. May, *Extinction Rates* (Oxford: Oxford University Press, 2005).
[18] C. Mora, D. Tittensor, A. Sina Adl, G. Simpson, and B. Worm, "How Many Species Are There on Earth and in the Ocean?" *PLOS Biology* (2011), https://doi.org/10.1371/journal.pbio.1001127.
[19] https://www.ipbes.net/news/million-threatened-species-thirteen-questions-answers#Q2.

and diversification rates, DeVos et al. estimate an actual rate of extinction of at least 1000 times normal background rates, and perhaps as high as 10,000× the background rate.[20] In a similar vein, based on a study of 29,400 vertebrate species, Ceballos et al. have calculated that 237,000 vertebrate species have become extinct since 1900.[21] This translates as roughly 1975 vertebrate species succumbing to extinction each year. This builds on earlier work of Ceballos and Ehrlich,[22] and since it deals with neither invertebrates nor plants, it suggests that the total rate of extinctions is well into mass extinction territory.

As a general rule, it is very difficult to find anyone now who disagrees with the claim that species extinctions are proceeding at a rate that is either hundreds or thousands of times faster than the normal or background rates that prevailed prehuman times during the last 10 million years or so.[23] We might, by arbitrarily selecting a high numerical threshold for mass extinction—10,000×, 20,000×, and so on—define it out of existence. But this would in no way alter the fact that something catastrophic is currently happening to the planet's species. In fact, an undue focus on extinction—especially global extinction—might prevent us from seeing clearly what this is.

Extinction: Beyond Antiquarianism

In the face of our bioexistential culpability, great solace can often be found in asking the wrong questions or focusing on the wrong things. Above all, we should not fall into the trap of thinking of extinction, and its disvalue, in what we might call *antiquarian* terms. The term "antiquarian" here is borrowed from Friedrich Nietzsche's essay, "On the Use and Abuse of History

[20] J. DeVos, L. Joppa, J. Gittleman, P. Stephens, and S. Pimm, "Estimating the Normal Background Rate of Species Extinction," *Conservation Biology* 29, no. 2 (2015): 452–462. This figure results, in part, from their much lower estimation of prehuman extinction rates as 0.1 extinctions per million species per year.

[21] G. Ceballos, P. Ehrlich, and P. Raven, "Vertebrates on the Brink as Indicators of Biological Annihilation and the Sixth Mass Extinction," *Proceedings of the National Academy of Sciences*, June 1, 2020, https://doi.org/10.1073/pnas.1922686117.

[22] G. Ceballos, P. Ehrlich, A. Barnosky, A. Garcia, R. Pringle, and T. Palmer, "Accelerated Modern Human–Induced Species Losses: Entering the Sixth Mass Extinction," *Science Advances* 1, no. 5 (2015): e1400253, doi:10.1126/sciadv.1400253.

[23] A. Barnosky, N. Matzke, S. Tomiya, G. Wogan, B. Swartz, T. Quental, C. Marshall, J. McGuire, E. Lindsey, K. Maguire, B. Mersey, and A. Ferrer, "Has the Earth's Sixth Mass Extinction Already Arrived?" *Nature* 471 (2011): 51–57.

for Life," in which he describes the *antiquarian bookseller*.[24] An antiquarian bookseller looks at his shelves of rare books and thinks, "Ah, yes, I have one of those!" Nietzsche's antiquarian bookseller is, actually, a collector of facts, divorced from the—historical and cultural—context that breathes life into them. The antiquarian collector of species, as I envision him or her, adopts the same attitude to his chosen passion. And having one of those, in this case, might amount to nothing more than the maintenance of a small, isolated population of animals of a given species with just enough genetic variation to sustain itself.

This is not to say that the antiquarian collection of species is unconcerned about decline in numbers. However, the antiquarian's guiding core value is the preservation of the species. A drastic decline in the numbers of a given species is important to the antiquarian only as a possible harbinger of that species' extinction. Thus, as long as breeding populations with sufficient genetic diversity to ensure indefinite continuation of the species are clearly maintained, with robust and long-term plans for this maintenance, this worry of the antiquarian species collector can be dispelled. It is not actual numbers of individual member of species, but only the nagging threat of extinction that troubles him.

This, I shall argue in the next chapter, is a warped way of understanding the harm we are currently doing to species. I shall argue that harm—existential harm—is done to a species when we drastically whittle down its numbers, and reduce its ancestral range, and this harm accrues whether or not extinction subsequently ensues. To see this, it is first necessary to understand what I shall call the *being* of a species. It is to this that we now turn.

---

[24] This is the second essay in Friedrich Nietzsche, *Untimely Meditations*. Cambridge Texts in the History of Philosophy, ed. D. Breazale, trans. R. Hollingdale (Cambridge: Cambridge University Press).

9

The Biomass Reallocation Program

Apocalypse Now

Imagine the mammoths of Wrangel Island lived on today. While you are at it, imagine that Wrangel Island is significantly bigger than it actually is, and the population of mammoths is large—say, 30,000 plus—and displays enough genetic diversity for the mammoths to flourish indefinitely. Having no natural predators—in particular, in this little thought experiment humans never arrived on Wrangel Island—the size of the mammoth population is constrained only by the availability of food and range. I am sure I am not alone in thinking that *this would be great!* However, it would be strange to conclude from this that, as far as the mammoth species is concerned, everything has come up smelling of roses.[1] On the contrary, the species would still have suffered a ruinous decline, and this, prima facie, seems to be a bad thing. A bad thing for mammoths. But how should we understand this badness? It can't be understood in terms of extinction—numbering 30,000 plus, and displaying sufficient genetic diversity, they are not threatened with extinction. And, in terms of their quality of life, there is no reason to suppose that each individual mammoth is any worse off than his or her prehistoric forebears. In what sense, then, is this imagined sharp decline in the numbers of mammoths a bad thing—for mammoths?

Imagine another postapocalyptic scenario, this one a little closer to home. The human race, let us imagine, has been reduced to localized remnants. Suppose, for example, a Hothouse Earth scenario unfolds, rendering much of the planet uninhabitable. The human species has largely been destroyed, but a few hardy survivors—numbering in the hundreds of thousands—manage to survive in the polar regions. This human population is stable, with enough genetic diversity to ensure its continuation. There is no threat of extinction in

---

[1] Which, not being seriously genetically compromised, these imagined mammoths might actually be able to smell. Fry et al. in "Functional Architecture of Deleterious Genetic Variants in the Genome of a Wrangel Island Mammoth," *Genome Biology and Evolution* 12, 2020, 48–58, have suggested that among the deficits of the mammoths of Wrangel Island was the loss of a sense of smell.

World on Fire. Mark Rowlands, Oxford University Press. © Oxford University Press 2021.
DOI: 10.1093/oso/9780197541890.003.0009

the near- to mid-term. But nor, given the parlous state of much of the rest of the world, is there any prospect of expanding human numbers much beyond these levels. Even though humans in this scenario are not threatened with extinction, it would be strange to conclude that all is well with the human race. On the contrary, it seems that a calamity has befallen them. But what is the nature of this calamity? Admittedly, the way I have set up the scenario, it sounds like humans are barely clinging to existence in the polar regions, which does seem a little grim. But we can rework the scenario to take this impression out of the picture. Suppose the polar regions in question have become transformed into idyllic places—veritable Gardens of Eden in the north and south of the planet. The humans who live there lead wonderful lives and, as a whole, are much happier than, say, the average human was in the first half of the twenty-first century. Being unaware that the human population was once much larger, they don't concern themselves worrying about the future of the human race, and things of that ilk. If we still want to assume that something unfortunate has happened to the species *Homo sapiens*—and, in effect, I defend that assumption in the pages to follow—then we can't understand this misfortune in terms of extinction, actual or threatened, and we can't understand it in terms of reduced quality of life. How, therefore, should we understand it?

Postapocalyptic scenarios don't have to be imagined. They are around us everywhere we look. They are just not *our* postapocalyptic scenarios. Wolves, in effect, are currently in their own postapocalyptic scenario—we humans were their apocalypse. At one time, gray wolves numbered in the tens of millions and ranged across most of the world. Today their numbers are estimated at 300,000, and they are confined to various pockets of the globe.[2] However, 300,000 is not extinction, in the antiquarian sense introduced at the end of the previous chapter—nowhere near it. At one time, wolves did look like they might become extinct, or something very close to it. But conservation efforts in different parts of the world have reaped rewards. The wolf is making a modest comeback in the contiguous 48 US states and in the mountainous and forested parts of Europe. Admittedly, in many regions of their comeback, wolves find themselves under pressure from hunters and ranchers. But suppose this were not so. Suppose, instead, that wolf numbers were naturally kept in check not through hunting and poisoning but by the

---

[2] D. Mech and L. Boitani, *Wolves: Behavior, Ecology and Conservation* (Chicago: University of Chicago Press, 2003), 231.

natural availability of food. As a result, global wolf numbers stabilized around the 300,000 mark. The probability or otherwise of this happening is not the point. My point is that lower wolf numbers do not necessarily translate into a lower quality of life for individual wolves. A reduction in wolf numbers of this magnitude is, intuitively—and I shall provide support for this intuition shortly—a *bad thing for wolves*. But the badness of it cannot be explained in terms of extinction and it cannot be explained in terms of deprivation of quality of life.

In these sorts of cases, it is tempting to suppose that a harm has befallen the respective species. But how do we understand this harm? It is not the simple reduction in numbers that is crucial. To see why, imagine another scenario. In this scenario, projections of human population growth turn out to have been catastrophically conservative, and by 2100 there are, let us suppose 50 billion people on the planet. In such circumstances, it would seem severe population reduction—in the region of magnitude of, perhaps, 80%–90%— would be beneficial. Not just for the planet, but for humans, too. Therefore, it seems reduction in the numbers of a species—even a severe reduction—is not, necessarily, harmful in itself. I shall develop an account that does justice to both intuitions: In some circumstances, a severe reduction in numbers of a given species is a bad thing for that species. But in other circumstances, a reduction of the same kind of magnitude is not a bad thing for that species. The reason for this, I shall argue, is that reduction in numbers is only a fallible indicator of the harm done to a species. Crucially, when harm is done to a species as its numbers are reduced, the harm done does not consist in the reduction in numbers itself but in something else. In the discussion that follows, I shall try to explain what.

In the previous chapter, I used the concept of *crime* to characterize human behavior toward other species. I also mentioned worrying about precisely *which* crime later on. That later on is now. The category of crime, I shall argue, is an *ontological crime*: a crime against the *being* of species. The expression, "ontological crime" was coined by Adriana Cavarero, in connection with certain crimes—murder, rape, torture, and so on—committed against humans.[3] An ontological crime, as Caravero defines it, is one performed with the intention of eliminating or distorting the being of a human individual. My use of the concept will be somewhat broader, obviously. But the general idea of

---

[3] A. Cavarero, *Horrorism: Naming Contemporary Violence*, trans. W. McCuaig (New York: Columbia University Press, 2009).

an assault on being will prove important. When we drastically reduce the numbers of a species, what we do may or may not lead to extinction. But extinction, if that is the result, is merely the culmination of a harm that starts much earlier, and of which the reduction in numbers is an indicator—albeit a fallible one. This harm, I shall argue, is best captured in terms of the idea of an assault on the *being* of a species.

The Value of Species

Many accounts of the harm that we do to species, including the harm of extinction, tacitly assume that the harm is really done to something else. Consider, for example, Aldo Leopold's famous ecological-moral principle: "A thing is right when it tends to preserve the integrity, stability, and beauty of the biotic community. It is wrong when it tends otherwise."[4] To the extent the loss of one of its constituent species would disrupt the integrity, stability, and/ or beauty of its surrounding "biotic community," this qualifies as "wrong"— that is, as a bad thing. This seems to make the value of species dependent on the value of biotic community—the surrounding ecosystem broadly construed. The biotic community is, fundamentally, the locus of value, and the value of the species derives from its role in promoting or sustaining the biotic community. In other words, the species itself has merely *instrumental* value: its value is that of an instrument. And the real—*intrinsic*—value lies with the biotic community.

One may ask further: why does the biotic community have value? And it's not long before things come back to us, self-absorbed species that we are. Biotic communities—ecosystems—are important because we humans depend on their health. The Intergovernmental Science-Policy Platform on Biodiversity and Ecosystem Services (IPBES) report mentioned in the previous chapter uses language of this sort: "Nature is essential for human existence and good quality of life. Most of nature's contributions to people are not fully replaceable, and some are irreplaceable."[5] Indeed, the entire report is framed in terms of what IPBES calls NCP—*Nature's Contribution to*

---

[4] A. Aldo Leopold, *A Sand County Almanac and Sketches Here and There* (New York: Oxford University Press, 1949), 224–225. Quoted in B. Callicott, "The Land Ethic," in *In Defense of the Land Ethic: Essays in Environmental Philosophy* (New York: SUNY Press, 1984), 204–217. Callicott's preferred version of this principle is: "A thing is right when it tends to disturb the biotic community only at normal spatial and temporal scales. It is wrong when it tends otherwise" (216).

[5] *The Global Assessment Report on Biodiversity and Ecosystem Services*, 12.

People. The message is clear: we should care about the health of ecosystems—and of species to the extent they contribute to this health—because healthy ecosystems are essential to human well-being. This is a pragmatic justification for preservation of species. Species and ecosystems have instrumental value. Their value lies only in their value *for us*.

Not all accounts of the value of species need be so nakedly pragmatic or instrumentalist. But it is difficult to rid them of their anthropocentrism. Leopold's foundational ecological-moral principle also mentions *beauty*. The idea that the value of species can be understood in aesthetic terms has been defended by Lilly Marlene-Russow: "Individual animals can have, to a greater or lesser degree, aesthetic value: they are valued for their simple beauty, for their awesomeness, for their intriguing adaptations, for their rarity and for many other reasons. We have moral obligations to protect things of aesthetic value."[6] We should protect species, on this view, because of the aesthetic value of their members. There was a strong suggestion of this kind of sentiment in the previous chapter, in my bewailing the loss of Pleistocene megafauna. On this view, the rarity of a species can enhance the aesthetic value of its members. Thus, the view can accommodate the idea that we have special obligations to species on the brink of extinction. However, aesthetic value does seem to be a somewhat uncertain foundation on which to build the value of species. Beauty, we are often told, is in the eye of the beholder—and that beholder seems to be us. This suggests that we could (legitimately!) stop worrying about the problem of mass extinction by refusing to develop the requisite aesthetic sensibility (or ruthlessly ridding ourselves of it if we already have it).

In this chapter, I shall not directly address the issue of value of species. Rather, my focus will be on the harm that is done to a species when its numbers are drastically reduced—as, for example, in the case of wolves today. Nevertheless, the view I shall propose can accommodate Leopold's emphasis on the contribution a species makes to the ecosystem in which it is embedded. It can also accommodate Russow's claims about the aesthetic value of species and their members. However, on my view, the harm done to a species when its numbers are drastically reduced is harm that is independent of the harm done to ecosystems and to aesthetic sensibilities. The harm is done, fundamentally, to the species. It is a harm done to the being of a species.

---

[6] L. Russow, "Why Do Species Matter?" *Environmental Ethics* 3 (1981): 101–112, at 112.

Species and Their Being

Anything that exists has *being*. This is a tautology: to have existence is to have being—at least according to the way I shall use these terms. Organisms and species exist, and therefore they have being in this sense.[7] But what does it mean to talk of the being of a species? The being of a species is a function—an extrapolation from and abstract description of—the being of its individual members. And whether we are trying to understand the being of an organism or species, it is perhaps easiest to start with the being of those organisms, and that species, we know best: us.

Each individual human is a biological organism, an instance of the genus *Homo* and species *sapiens*, constructed from a blueprint supplied by a specific genetic profile that we have now been able to map. However, now consider another question: What is the *being* of a human? Here is another way of putting this question: What type of *existence* does a human being have? Or, perhaps, a slight improvement on this: *In what ways does a human exist?* A human might be a biological organism, but the ways in which a human exists are not biological organisms.[8] Thus, on the one hand, there is a human being: a biological organism. On the other hand, there are the ways in which this biological organism exists.

I am going to call the latter—the ways in which a human exists—the *being* of the human. Being, in this sense, is something had both by individual human organisms and also by humans taken as a species. The being of humans, understood as a species, will be an abstract description of the being of individual humans, broad enough to capture and subsume the ways in which different humans might exist. This way of thinking about being derives from Martin Heidegger's *Being and Time*.[9] This book is an inquiry into the being of humans, an initial stage in a larger project—which he never completed—of understanding the idea of being itself. Heidegger also thought that the best way to understand the idea of being in general was to inquire about the being of the beings we know best: humans. Any given human is

---

[7] I am aware, of course, that there is a debate about the reality of species—a debate in which I shall determinedly refuse to become involved. I do not think the standard arguments against the reality of species are very good anyway. And, given that this chapter, and the previous one, is about the problem of mass extinction of *species*, it would be difficult to worry about this if there were, in fact, no such thing as species.

[8] To confuse individual humans with the ways they exist would be to commit a *category mistake* in the sense made famous by Gilbert Ryle in *The Concept of Mind* (London: Hutchinson, 1949).

[9] Martin Heidegger, *Being and Time*, trans., J. Maquarrie and E. Robinson (Oxford: Blackwell, 1927/1962).

a biological organism that takes up a certain region of space and follows a singular track, in principle identifiable, through space and time. The being of that human, however, is not like this at all. The being of a human is, very roughly, what humans characteristically do and the way they go about doing it. Of course, we do many things. As I type these words, I am, for example, about to go to the supermarket. But what Heidegger was looking for was a far more abstract description, true of all humans.

This being—way of existing—of humans, Heidegger labeled *dasein*. *Dasein*, or *da-sein*, translates as *being-there*. Heidegger provided a detailed analysis of *dasein* as, wait for it: *already-being-in-the-world-toward-a-future-with others*. I am a *being-toward-a-future* in that I seek to bring about certain futures rather than others—in effect, to transform goals into facts. I, for example, want to finish writing this book by the end of the summer. At present, it is a goal. By the end of the summer I hope it is a fact that I have finished the book. That is one of my more pressing ways of being-toward-a-future. This goal-seeking—or future-determining—behavior of mine does not spring from nothing or nowhere. At any given moment in time, I always find myself in a situation of *already being-in-the-world*. The goal of finishing the book by the end of the summer springs from my, rather carelessly I have to admit, allowing myself to be appointed—or is that, perhaps, anointed—chair of my department at a rather challenging time in the history of higher education. As a result, I suspect, once August comes around, that is going to be it for research and writing for the next 9 months. My being-toward-a-future and my already-being-in-the-world are, in this way, interrelated. In particular, the urgency with which I pursue this goal is a function of my situation—my already-being-in-the-world.

I also mentioned that I soon have to go to the grocery store. This is because my sons are hungry, and they insist that there is no suitable food in the house and therefore I must go on what they call, in this time of Corona, a *suicide run* (mockingly, of course). I am, in this sense, *being-with-others*, and this is closely entwined with my being-toward-a-future and my already-being-in-the-world. It is my being-with-my-sons that helps determine my already-being-in-the-world and my being-toward-a-future. My already-being-in-the-world, in this case, consists in a situation I find myself in: having to provide for two, seemingly always hungry, boys. This helps determine this aspect of my being-toward-a-future: I have to go to the store because I exist with sons who are hungry (part of *their* already-being-in-the-world) and my *concern* for them—which is part of the way I am *with* them—dictates that my immediate future will take on the form of going to the grocery store. I am, in this sense, *already-being-in-the-world-toward-a-future-with-others*. I am a biological organism. But *already-being-in-the-world-toward-a-future-with others* is the type of being that I have. The same is true of all other humans. This characterization—*already-being-in-the-world-toward-a-future-with-others*—is, Heidegger argued, sufficiently broad and abstract to qualify as a description of the being of humans in general.

That is more than enough Heidegger for any one day. You don't have to believe his analysis of the being of humans. What I want you to take away from Heidegger is not his specific analysis of human being-in-the-world but, rather, his distinction between humans as biological organisms and the type of being had by humans—roughly, what they do and the way they go about doing it. Humans aim for a future, guided by a situation in which they are already in and by the others around them for whom they care (by loving them, hating them, feeling obligated to them, being indifferent to them, and many other ways besides). Understood as an individual organism, a human has fairly clearly defined boundaries, which more or less coincide with the skin. But the being of humans—what they do and the way they go about doing it—is not like this at all. The being of any given human is spread out, *distributed* onto the world. You cannot precisely locate what a human does and the way he or she goes about doing it. It is wherever the *doing* is. When I go to the grocery store, this going to the store encompasses me, the route to the store, and store itself. More generally, this being—or doing—is spread out: it encompasses the world which I am already in, the future to which I aspire, the others which are indispensable parts of where I am and where I want to go, and the instrumental means (laptops, cars, grocery stores, etc.) that allow me to get there. It is this general distinction between an organism and its being that is important for the purposes of this chapter, rather than Heidegger's specific articulation of it.

Heidegger's analysis was intended as a general one, applicable to all humans. If the analysis is correct, therefore, Heidegger has provided an account of the *being* of humans in general. The *being of a species* is the characteristic way members of a given species do what they do and the way they go about doing it. But, while a general analysis, Heidegger's account is also, in a clear sense, rather parochial. There is *a lot* he didn't mention. Each one of us humans does many things not incorporated into Heidegger's analysis. He didn't mention respiration, for example, which, when you think about it, is really a rather important thing that humans do and, accordingly, something of which I like to make a habit. Heidegger's interests lay elsewhere. He was engaged in what he once called *philosophical anthropology*. There is absolutely no need to concern yourself with what that is. My point is that what features of the being of an organism or species are included in one's analysis depends on the interest one has in conducting the analysis and, in particular, on the theoretical perspective from which one constructs the analysis. The theoretical perspective that is germane to the issue of species, and the harm done to them, is the perspective provided by the science of ecology. What I am going to argue next is that there is a definite, and quite specific, conception of the being of any species that is built into this science. The next section is concerned with identifying this conception of being.

The Dasein of a Wolf

The well-known, and very effectively popularized, return of wolves to the Greater Yellowstone ecosystem provides a graphic, and therefore very useful, illustration of the being of wolves as this being is conceptualized in the science of ecology.[10] Wolves became extinct in the Great Yellowstone ecosystem in 1926, victims of the general policy of extermination pursued by successive US administrations. They were reintroduced in 1995—with some quite startling results.

In the absence of this keystone predator, the numbers of deer and elk—despite repeated attempts by humans to limit their populations—had exploded. As a result, vegetation in the ecosystem became severely degraded. The reintroduced wolves quickly changed this. In part, this was because they began to trim deer and elk numbers. However, the most important factor was not the reduction in their numbers of deer and elk, but a *change in their behavior* induced by the wolves. In particular, deer and elk could no longer afford to relax in terrain in which they might easily become trapped—most notably, river valleys and gorges. If you absolutely, positively have to enter a river valley, when you are a deer and there are wolves around, then the order of

---

[10] For a popularized version, see, for example, *How Wolves Change Rivers*, narrated by George Monbiot, https://www.youtube.com/watch?v=ysa5OBhXz-Q. For some peer-reviewed work underlying this, see D. Fortin, H. Beyer, M. Boyce, D. Smith, S. Duchesne, and J. Mao, "Wolves Influence Elk Movements: Behavior Shapes a Trophic Cascade in Yellowstone National Park," *Ecology* 86, no. 5 (2005): 1320–1330. Also see J. Laundre, L. Hernandez, and K. Altendorf, "Wolves, Elk, and Bison: Reestablishing the 'Landscape of Fear' in Yellowstone National Park, U.S.A.," *Canadian Journal of Zoology* 79, no. 9 (2001): 1401–1409.

the day is: get in and get out as fast as you can. As a result, vegetation in these places—birches, aspens, willows, and so on—quickly recovered. The return of the trees saw the return of animals that depend on trees—most notably, birds and beavers. Beavers, of course, build dams, which create lakes, which provided suitable habitat for other creatures, such as otters and muskrats, fish and ducks, which also began to return to the area in increasing numbers. The wolves killed coyotes, which allowed rabbits, mice, badgers, and weasels to proliferate which, in turn, caused the numbers of eagles, hawks, and ravens to increase. Bear populations also increased, because of the increase in carrion and newly available berries on the regenerating shrubs. Most strikingly of all, the wolves changed the way in which the rivers of Yellowstone flowed. The regenerating forests stabilized the banks and reduced soil erosion. Rivers became more fixed in their paths, meandering less, alternating between pools and swiftly flowing narrower channels.

These effects, rippling down from the reintroduction of the wolf to the Greater Yellowstone ecosystem—ripples extending to deer, elk, beavers, birds, coyotes, bears, rivers, and more—is known as a *trophic cascade*. The wolf's place in this ecosystem is, in effect, defined and demarcated by its place or role in this cascade of rippling effects. Of course, one might question this *specific* analysis of the wolf's role in its Greater Yellowstone ecosystem. But what is important for my purposes is that the wolf's role in the ecosystem can be defined in these sorts of terms. The reason this is important is that the same claim can be translated into the language of *being*. From the perspective of the science of ecology, *what a wolf does and the way it does it* is delimited by this kind of trophic cascade that it produces in its ecosystem. This rippling cascade, therefore, is the *being* of the wolf as this is understood from the perspective supplied by the science of ecology.

The being of the wolf is, on this conception, a relational one: it incorporates the wolf's relation to other animals, to vegetation, and to the land more generally. In different ecosystems, the cascade that emanates from, and centers around, the wolf will be somewhat different. There is little reason to suppose that the cascade produced by the timber wolf in the Greater Yellowstone ecosystem will be identical with that produced by, for example, the Arabian wolf in the deserts and marshes of southwestern Iraq.

Being a keystone predator, the trophic cascade that, in any given ecosystem, makes up the being of a wolf is a relatively dramatic one, and therefore quite easy to identify, at least in broad outline. But the foundational idea of the science of ecology is that each species will have its own *what it does and the way it does it*—its own *being*—and this corresponds to its role in its ecosystem to which it has become coadapted. Thus, these kinds of trophic ripples, radiating outward from any member of any given species, connecting it with, anchoring it to, its ancestral home, and to the other residents of that home, constitute the *being* of that species. When we exterminated all the wolves of the greater Yellowstone ecosystem, we ripped from the fabric of that ecosystem these rippling trophic cascades that made up the being of the wolf. In this sense, the extermination of the wolf was an assault on the being of this species. Once we see clearly the role that the wolf plays in the Greater Yellowstone ecosystem—its place in a system of rippling trophic cascades—we can also appreciate just how much it is now *not* playing this role in other places, many of them not unlike Yellowstone, where it might have played this kind of role and once did so. The being of the wolf is progressively denuded. More and more this being becomes parceled into smaller and smaller pockets—shrinking reservations of being. The wolf has not become extinct yet, but this damage to its being has already been done.

One advantage of this account is that it explains why a reduction in numbers of the same kind of magnitude would not count as a harm in the circumstances of severe overpopulation. In such circumstances, the being of a species has become warped or distorted. From the perspective provided by the science of ecology—which focuses on the developmental integrity of ecosystems over normal spatial parameters[11]—the notion of the being of a species is a *normative* one. It is defined by the role the species is *supposed* to play in such a system—by the rippling trophic cascades that are *supposed* to radiate outward from it.[12] If circumstances of overpopulation prevent a species from playing this role in a given ecosystem, then the being of that species isn't, at that time, in place. In these circumstances, a reduction of its numbers would not count as an assault on the being of this species.

The appeal to the integrity of ecosystems—or biotic communities—brings to mind Leopold's view: "A thing is right when it tends to preserve the integrity, stability, and beauty of the biotic community. It is wrong

---

[11] The qualification "over normal spatial parameters" is important, and it is introduced to accommodate the fact that ecosystems are always evolving. What is key to the health of an ecosystem is, of course, that changes do not occur too quickly.

[12] Ruth Millikan would call this the "relational proper function" of the deer. See her *Language, Thought and Other Biological Categories* (Cambridge, MA: MIT Press, 1984). My suggestion, in effect, is that the being of a species can be understood in terms of the concept of relational proper function. Note there is nothing in my account that entails a static conception of the being of a species. The being of a species can change over time because ecosystems can, and typically do, change over time.

when it tends otherwise." However, the view I have defended is subtly different. On Leopold's view, when the numbers of a species are drastically reduced, the primary harm is done to an ecosystem, a harm that is perpetrated via an assault on its component species. Of course, in such circumstances harm is, or may be, done to an ecosystem. However, on my view, harm is also done to the species. The being of the species is under assault. The integrity of the ecosystem over normal temporal parameters may define, or normatively constitute, when a harm is done to the being of a species. But the harm is, nevertheless, done to the species through an assault on its being.

This account is also able to incorporate both pragmatic and aesthetic considerations. Some, for example, may find the rippling trophic cascade that constitutes the ecological being of the wolf beautiful. I do. In fact, I have difficult conceiving of anything more beautiful. Others, of course, may be unmoved by this. Once we realize the inextricable entanglement of the being of the wolf with the Greater Yellowstone ecosystem, we may realize how pragmatically foolish it was to remove the wolf from so much of its ancestral land. But what unites and underlies these considerations is the idea of the being of the wolf: the what the wolf does and the way it does it in the ecosystem that is its ancestral home. The being of the wolf species consists in the rippling trophic cascades that center around it. When we rip these away from its ancestral home, from places like the Greater Yellowstone ecosystem, we are engaged in an assault on the being of the wolf. The assault is real, and pernicious, whether or not it culminates in extinction.

Biomass Reallocation and the Assault on Being

A graphic indicator of our human assault on the being of many species can be found in the *redistribution of biomass* we have orchestrated during our tenure on this planet. What makes biomass an especially useful indicator of being, in the sense outlined earlier, is that, like "being," "biomass" is a *mass* term, as opposed to a *count* term, such as "organism." Mass terms refer to "stuff"—quantities and qualities broadly construed. Count terms refer to objects.[13] The stuff to which mass terms denote come in greater or

---

[13] For a classic study, see J. Pelletier, *Mass Terms: Some Philosophical Problems* (New York: Springer, 1979).

lesser amounts. The object to which count terms refer come in greater or lesser numbers.

The best efforts of COVID-19 notwithstanding, we are, as a species, still in astonishingly rude health. There are just so many of us—closing in quite quickly on 8 billion, as I write. Projections have our population increasing to 11 billion around the turn of the next century. And we are absolutely everywhere! But there are several other species that are also in remarkably good health: the species of animals we eat. There are billions of them, too—they outnumber us by a margin of three to one—and because we are almost everywhere, they are almost everywhere, too.[14] Unfortunately, the same can't be said for most other species.

However, this is all numbers. Let's switch from talk of numbers to biomass. A biomass calculation for a species, or other group, is the average mass of each individual of the species multiplied by the number of individuals in that species or group. As we saw in Chapter 1, a biomass calculation for all mammals yields some truly disturbing figures. According to the calculations of Bar-On, Phillips, and Milo, livestock make up 60% of all mammalian biomass.[15] Humans make up 36%. This leaves a mere 4% of mammalian biomass for all other mammals. That is, humans and the mammals we eat make up 96% of all mammalian biomass! To wild mammals we have allotted a measly 4%. Almost all of the terrestrial mammalian biomass on planet earth today is made up of humans and the mammals humans eat. In the oceans, centuries of whaling have left only one-fifth of the marine mammalian biomass that there once was. Since our Paleolithic dawn, we humans have destroyed nearly 85% of wild terrestrial mammalian biomass, and 80% of wild marine mammalian biomass.

For birds, the picture is no less bleak. Bar-On, Phillips, and Milo estimate that, today, 70% of avian biomass on earth is made up of domesticated birds—chicken and other poultry. Only 30% of avian biomass belongs to wild birds. That sounds like an improvement only until we remember that human biomass is not factored into this figure as it was in the case of mammals. Measured in terms of biomass, almost all mammals on Planet Earth are either humans or mammals that humans eat. A sizeable majority of birds are

---

[14] For data, see United Nations, Food and Agriculture Organization, http://www.fao.org/faostat/en/#data/QA.

[15] Y. Bar-On, R. Phillips, and R. Milo, "The Biomass Distribution on Earth," *Proceedings of the National Academy of Sciences* 115 (2018): 6506–6511. https://doi.org/10.1073/pnas.1711842115.

domestic ones, raised for their meat or their eggs. The picture is stunning. Depressing. And hardly unexpected.

At an abstract level of description, one might think of us humans as being engaged in a massive *biomass reallocation program*. Before our dramatic, Paleolithic entrance onto the world stage, the biomass of mammals and birds was distributed relatively evenly, allocated widely among the barely imaginable beasts we talked about in the previous chapter, and many others besides, including us. Our early efforts were as much reductive as redistributive. We reduced mammalian and avian biomass wherever we encountered it, converting it into human biomass. Our numbers grew, and those of other animals declined, to the point of extinction in many cases. Then, as the Paleolithic age turned into its Neolithic successor, many of us became farmers, and the program of biomass reallocation moved into full swing. Biomass formerly allocated to wild creatures now became biomass allocated to domestic livestock and fowl—and through them, ultimately, to us.

As we have seen, this program was astonishingly successful. The vast majority of mammalian and avian biomass on the planet is either us or biomass that we use to build and maintain our own biomass. The biomass of everything else has dwindled catastrophically, almost, in the case of wild mammals, to the point of nothing. This remaining biomass exists in far-flung and scanty pools, collecting in the nonhuman interstices of the planet; on little islands of land that we haven't yet got around to converting to grow the biomass we consume. As for this remaining, remnant biomass of wild mammals and birds—we hunt it whenever we can. The world is ours and— almost—everything that's in it.

It seems that we have managed to preside over a staggering, barely comprehensible reduction in the biomass of wild mammals, terrestrial and marine, and birds. The grotesquely warped, skewed distribution of mammalian and avian biomass on the planet is a stark indicator of to what extent *what species do and the way they do it* has been ripped out of the fabric of this planet. The redistribution of biomass reflects an underlying assault on the being of species. This assault takes away from the planet the rippling, cascading trophic waves that bind it together. Even if no species had ever gone extinct, it is difficult to see how this assault could be anything other than a criminal act—a crime of an ontological variety. Extinction is, in effect, merely the final act in a crime. Sometimes we get around to it, sometimes not. But either way, a crime is in the process of commission. This is an ontological crime; a crime against being.

Drivers of Reallocation-Extinction

The reallocation of biomass is, at the same time, an assault on the being of species, one that sometimes, and increasingly often, culminates in extinction. Extinction, when it occurs, is the end product of a protracted process of biomass reallocation: the *reallocation-extinction* process. Harm occurs long before extinction occurs, if it occurs, and consists in an assault on the being of species.

The IPBES report on extinction lists, in order of importance, what they call *drivers of extinction*. All of them in, one way or another, are also ways through which biomass is reallocated, and so we can, with equal justification, regard them as drivers of biomass reallocation. Whether of extinction or reallocation, these drivers have increased dramatically in force and scope during the past 50 years. The direct drivers, in order of importance, are (1) changes in land use and sea use, (2) direct exploitation of organisms, (3) climate change, (4) pollution, and (5) invasion of alien species.[16] Indirect drivers of reallocation-extinction are, essentially, those that are ultimately responsible for the direct drivers. These include "production and consumption patterns, human population dynamics and trends, trade, technological innovations and local through global governance."[17]

The two main causes of the biomass reallocation-species extinction process, therefore, are changes in land and sea use, and direct exploitation of organisms. The expression *direct exploitation* refers to the hunting of animals—terrestrial or marine—for their meat or other products. In the case of marine animals, (1) and (2) are reversed: fishing is the primary cause of marine species extinction. On land, however, change in land use is the primary driver of reallocation-extinction, followed by hunting. Change in land use can be the result of several factors. Urban development and the expansion of infrastructure linked to a growing population and increased consumption can obviously be one of those factors. However, it is not the

[16] *Global Assessment Report on Biodiversity and Ecosystem Services*, https://ipbes.net/global-assessment.
[17] *Global Assessment Report on Biodiversity and Ecosystem Services*, https://ipbes.net/sites/default/files/ipbes_global_assessment_chapter_2_1_drivers_unedited_31may.pdf.

most important. That accolade belongs to agriculture. The IPBES report tells us that "Agricultural expansion is the most widespread form of land-use change, with over one third of the terrestrial land surface being used for cropping or animal husbandry."[18] The report also notes that this change in land use has come "mostly at the expense of forests (largely old-growth, tropical forests), wetlands and grasslands."[19] Thus, to summarize: on land, the number-one driver of species extinction lies in changes in land use, and the number-one reason for change in land use is agricultural expansion. The number-one driver of reallocation-extinction on land, therefore, is agricultural expansion.

Agriculture is, of course, both arable and pastoral. However, pastoral farming utilizes far more land, and most change in land use is, accordingly, driven by pastoral farming. Ground Zero for change in land use is, almost certainly, Brazil. In the Amazon rainforest, according to the Satellite Legal Amazon Deforestation Monitoring Project, almost 10,000 square kilometers, or 3800 square miles, or 2,432,000 acres were lost to deforestation in the year August 2018 to August 2019.[20] Seventy percent of this deforested land is used as pasture, with feed crop cultivation occupying much of the remainder.[21]

The conclusion seems inescapable. If, as IPBES tells us, (1) changes in land use are the number-one factor in driving the reallocation-extinction process, and if, as IPBES also mentions, (2) the number-one factor driving changes in land use is agricultural expansion, and if (3) by far the predominant form of agricultural expansion is in pastoral or animal farming, together with the growing of feed crops to support this animal farming, then we can conclude that (4) the most important factor driving the biomass reallocation and subsequent extinction of terrestrial species is the expansion in animal agriculture.

The tenure of humans on this planet has been dominated and defined by a vast biomass reallocation program that, in effect, amounts to an assault on the being of so many species. This reallocation initially took the form of hunting, which drove the Quaternary extinction event. Today, it is farming—and, in particular, the farming of animals—that is the most significant factor

---

[18] *Global Assessment Report on Biodiversity and Ecosystem Services*, https://ipbes.net/global-assessment, 14.

[19] *Global Assessment Report on Biodiversity and Ecosystem Services*, https://ipbes.net/global-assessment, 13.

[20] CGOBT/INPE, Earth Observation, http://www.obt.inpe.br/OBT.

[21] Food and Agriculture Organization (FAO) of the United Nations, http://www.fao.org/publications/sofa/2012/en/.

in continuing this reallocation. However, the common factor uniting both hunting and pastoral farming (not to mention the majority of arable farming which is used to grow animal feed crops) is what these are for: meat. Eating meat is a habit, not a necessity. If we could break this habit then, in conjunction with the kind of afforestation program required to address the problem of climate change, we could reverse the distribution of terrestrial mammal biomass. Biomass allocated to the mammals we currently eat could return to some of its more ancestral vehicles: wolves, bear, lions, tigers, panthers, elephants, eagles, and myriad others—all the magnificent creatures we didn't quite get around to killing during our Quaternary killing days and subsequent mopping-up efforts. The land could be remade into a suitable home for these animals—the transition from cropland to forest is another kind of biomass reallocation. At the same time, the oceans would be able to replenish their devastated biomass.

We are the authors of, and are presiding over, another *great dying*. And, in reality, we always have been. This great dying is, fundamentally, an assault on the being of species. Before it takes a species, the assault first strips away its being. It rips being from the world, wherever it can, the *what a species does and the way it does it*. This kind of assault eventually translates, in many cases, into extinction. But the harm is there and real long before extinction occurs. There are many, many species of animals and plants today that, even though they have not yet become extinct, are effectively in circumstances not too dissimilar from the last mammoths of Wrangel Island. Depleted, decimated: tiny pockets of being confined to tiny pockets of their ancestral lands. Waiting for extinction. There is, today, a great dying of the earth's animals and plants. And the primary reason for this—the reason against which all others pale in comparison—is obvious: we eat animals. We eat animals and we need to stop.

10

Pale Horse

Locked Down!

In my view, what this book has been lacking in the past nine chapters or so—
you might have your own ideas on this, but what *I* think has been lacking—
is a good old dispute of the Biblical interpretation variety. Take the *Four
Horsemen of the Apocalypse.* Conquest, War, Famine, and Death, some people
say—and that does seem to be what *Revelation* 6 says. But Conquest and War
seem to be pretty much the same thing. Granted, War might be a more general category than Conquest, assuming not all wars are wars of conquest.
But, still, a bit too much overlap really. The Four Somewhat Overlapping
Horsemen of the Apocalypse? Doesn't sound quite as menacing. By the
time we get to 6:8 things have become even more confusing. The *King James
Version*—obviously the best version of the Bible—tells us: "And I looked,
and behold a pale horse: and his name that sat on him was Death, and Hell
followed with him. And power was given unto them over the fourth part of
the earth, to kill with sword, and with hunger, and with death, and with the
beasts of the earth." But this makes it seem as if there was one horseman,
death, who was able to delegate authority—to the minions of Hell, perhaps?—to kill by one of four means. Sword—okay, that's war. Hunger—okay,
that's famine. But with death? How do you kill someone with death? I can
kill you with a sword or by starving you to death. But kill you with death?
Death seems to be more the result of killing someone, rather than something
you kill them *with*. A category mistake, as Gilbert Ryle would have called it.
Beasts of the earth, though. That is interesting. I'll get back to you on that.

The *English Standard Version* is somewhat different: "And I looked, and behold, a pale horse! And its rider's name was Death, and Hades followed him.
And they were given authority over a fourth of the earth, to kill with sword
and with famine and with pestilence and by wild beasts of the earth." Killed
with pestilence, instead of killed with death. That does make more sense.
Nevertheless, puzzles remain. The way this passage is written makes Death
the central figure. He gets a horse, then. But sword, famine, pestilence, and

World on Fire. Mark Rowlands, Oxford University Press. © Oxford University Press 2021.
DOI: 10.1093/oso/9780197541890.003.0010

the wild beasts of the earth seem merely means by which death is brought about. It's not clear why they would get a horse at all. Ponies perhaps. But not a horse like death. Conversely, if Death has a pale horse, and sword, famine, pestilence, and the wild beasts of the earth also have horses, then shouldn't there be five horsemen of the apocalypse?

Some versions contest the color of Death's horse. Both the *Christian Standard Bible* and the *Contemporary English Version* tell us that Death, in fact, sat on a "pale green horse." What, like Benjamin Moore, Silken Pine, 2144-50? My wife wanted to lockdown-paint the living room that color a few weeks ago. Whoa, I said, a bit four horseman-y isn't it? What about a nice, cheerful yellow? Benjamin Moore AF330, or something like that? As far as I am aware, none of the horses were that color. She will probably leave me soon. Curiously, Death—which according to 6:8, definitely seems to be the main man/woman/nonbinary figure in the apocalypse—sometimes seems to have to share his/her/their horse with pestilence, who is also said to ride a pale horse. At least, according to the *Encyclopedia Britannica*, and who am I to argue with that?[1] Maybe death has a pale green horse, and pestilence just a run-of-the-mill pale horse? I don't know. All this shows, of course, is that I have been cooped up far too long. I need to get out more.

As I write these words of shoddy Biblical interpretation, the world has been grappling with COVID-19, the disease that results from the SARS-CoV-2 virus. No one alive—apart from the rare centenarian survivor of the Spanish flu—has seen anything like this before. Many countries of the world are just beginning to emerge from a lockdown that has lasted several months. And, as a result, case numbers in some of those countries are already spiraling upward again. The official global death toll stands, today, at just over 700,000.[2] Initial excess death estimates, however, suggest a much higher figure.[3] For some survivors, certain long-term, life-changing, health consequences of the virus are expected—including pulmonary fibrosis, or irreversible scarring of the lungs, and also possible neurological damage.[4] The economic

---

[1] *Encyclopedia Britannica*, https://www.britannica.com/topic/four-horsemen-of-the-Apocalypse.

[2] https://www.worldometers.info/coronavirus/. Accessed August 4, 2020.

[3] https://www.cdc.gov/mmwr/volumes/69/wr/mm6919e5.htm. May 15, 2020 / 69(19);603–605. For more details, including variability from one country to another, see *The Economist*, "Tracking Covid-19 Excess Death across Different Countries," https://www.economist.com/graphic-detail/2020/04/16/tracking-covid-19-excess-deaths-across-countries. Accessed June 23, 2020.

[4] At least, according to leaked NHS (National Health Service)guidelines seen by the *Daily Telegraph*. Laura Donnelly and Victoria Ward "Revealed: Scars of Covid-19 Could Last for Life as Doctors Warn of Long-Term Damage to Health," https://www.telegraph.co.uk/news/2020/06/22/revealed-scars-covid-19-could-last-life-doctors-warn-long-term/. June 22, 2020.

See also Zoe Cormier, "How Covid-19 can damage the brain" https://www.bbc.com/future/article/20200622-the-long-term-effects-of-covid-19-infection.

consequences have been unprecedented and devastating—with unem-ployment in the West reaching levels not seen since the Great Depression. Prospects of the development a viable vaccine are, as yet, guarded. This is, in part, because we don't yet understand the nature of the immune response through which people have successfully fought this virus. Some people do not seem to develop antibodies. And in those who do, these antibodies only seem to last a few months. Even if a vaccine is developed, the worry is that the people it will help least are the very people who need it most—those over 70.

The truly worrying thing about COVID-19 is the suspicion that it is far from the worst disease that might one day engage us. The consequences described in the previous paragraph are consequences of a disease whose mortality rate seems to be in the region of between 0.5% and 1%.[5] One can only imagine the consequences of a disease with a mortality rate of 10% or higher. As we shall see, there are some on the horizon. The good news is that COVID-19 was not unavoidable. There is one, and only one, real reason why COVID-19 exists in humans. The devastation it has visited on the planet's human population is a consequence of one thing only: *we eat animals*. The bad news is that, if we continue to do so, COVID-19 won't be the last, and is unlikely to be the worst, disease that we will be forced to face.

Zoonoses: A Brief Primer

A *zoonosis* is an infectious disease (caused by viruses, bacteria, or parasites) that has spread from a vertebrate to humans. The animal that originally passes on the disease is known as the *reservoir*. Sometimes a disease that is passed from animals to humans can mutate to allow transmission from human to human. The more efficient the transmission between humans, the less important the role of the animal reservoir becomes. Eventually, the dis-ease might become an exclusively human one. If this happens, the disease no longer technically qualifies as a zoonosis but, rather, is said to have a *zoo-notic origin*.

Any *novel* pathogen must, of course, come from somewhere. By definition, the reservoir for a novel pathogen can't be humans—otherwise it wouldn't be

---

[5] *Journal of the American Medical Association, JAMA Intern Med.* 2020;180(8):1045–1046. https://jamanetwork.com/journals/jamainternalmedicine/fullarticle/2766121?guestAccessKey= 28f9727c-0a86-4d26-8900-e4391ec18af2&utm_source=silverchair&utm_medium=email&utm_ campaign=article_alert-jamainternalmedicine&utm_content=olf&utm_term=0051420.

novel. A sizeable majority of novel pathogens come from other vertebrates—they are zoonoses or have a zoonotic origin.[6] The same is true of most of the established human disease-causing pathogens—any established human pathogen was once a novel one, and its source was, in most cases, some vertebrate animal or other.[7] The kind of animal in question may be wild or domestic, and the distinction between wild and domestic animals loosely tracks the distinction between tropical and temperate diseases. Generally, as an imperfect rule of thumb with several exceptions, tropical diseases have their origin in wild animals, and temperate diseases tend to originate in domestic animals.[8]

The domestication of animals that began with the Neolithic farming revolution, in effect, brought humans into contact with an entirely new array of pathogens, triggering a mass transmission of animal diseases into human populations. The animals best suited for domestication are herd animals and, unfortunately, their habit of "herding" together in numbers also makes them the types of animals best suited to harbor epidemic diseases.[9] Let's take a look at some of the humanity's biggest killers, beginning with temperate diseases.

Tuberculosis is thought to have to have been acquired through the domestication of goats.[10] During the twentieth century alone it killed approximately 100 million people, and even today still kills an estimated 1 million people per year.[11] Measles is thought to have been acquired from domesticated sheep and goats, a mutant form of the *rinderpest* virus.[12] Although in the West, we tend to regard it as a relatively mild disease, it nevertheless kills upward of 100,000 people in any given year, mostly in sub-Saharan Africa. In 2018, according to the World Health Organization (WHO), measles killed

---

[6] As *The Lancet* puts it: "All human diseases to emerge in the last twenty years have had an animal source." See "Editorial. Avian Influenza: The Threat Looms," *Lancet* 363, no. 9405 (2004): 257.

[7] The alternative is *co-speciation* where an ancestral pathogen that was already present in the common ancestor of humans and other apes co-speciated when the human lineage diverged, c. 5 million years ago. *Falciparum malaria* might have become a human pathogen by this route. Of course, this just means that the pathogen was novel to our common ancestor.

[8] N. Wolfe, C. Dunavan, and J. Diamond, "Origins of Major Human Infectious Diseases," *Nature* 447 (2007): 279–283.

[9] Michael Greger, *How to Survive a Pandemic* (New York: Flatiron Books, 2020), 108.

[10] Espinosa de los Monteros, L., Galan, J., Guitierrez, M. et al., "Allele-Specific PCR Method Based on pncA and oxyR Sequences for Distinguishing Mycobacterium bovis from Mycobacterium tuberculosis: Intraspecific M. bovis pncA Sequence Polymorphism," *Journal of Clinical Microbiology* 36, no. 1 (1998): 239–242. Bovine tuberculosis appeared much later.

[11] World Health Organization, 2019, Global Health Observatory Data. "How Many TB Cases and Deaths Are There?" https: www.who.int/gho/tb/epidemic/cases_deaths/en/. Accessed June 23, 2020.

[12] Greger, *How to Survive a Pandemic*, 110.

140,000 people, mostly children under 5.[13] Over the last 150 years or so, it is estimated to have killed around 200 million people.[14] Diphtheria came from domesticated herbivores. Mumps probably came from domesticated pigs. Influenza A outbreaks come variously from domesticated ducks, geese, chickens, and pigs, and ultimately from wild ducks and, according to the WHO, tends to kill up 650,000 globally each year.[15] Smallpox is believed to have originated in camels. It is now eradicated, but it did much damage prior to this. In the twentieth century alone, it is estimated to have resulted in 300–500 million deaths.[16]

Helicobacter pylori is a type of bacterium that is thought to infect roughly half of the world's population. It is the cause of the vast majority of peptic ulcers in humans. It also causes stomach cancer. *H. pylori* is thought to have originated in sheep's milk.[17] Plague—*Yersinia pestis*, coming in bubonic, pneumonic, and septicemic forms—was spread to us from infected rodents via their fleas. In Europe, this came in separate, successive waves between the 1300s and 1600s. The wave of the 1300s is estimated to have killed up to half of the population of Europe.[18] Leprosy is thought to have originated in water buffalo.[19] Diphtheria, the feared killer of children—it killed the little boy who would have become my uncle—with mortality rates as high as 40% before the development of a vaccine, is thought to have originated in domestic herbivores.[20] Far less dramatic, but very persistent and irritating, the common cold is thought to have originated either in horses[21] or cattle.[22] With the exception of plague, these diseases were all acquired from animals that had been domesticated.

---

[13] World Health Organization, https://www.who.int/news-room/detail/05-12-2019-more-than-140-000-die-from-measles-as-cases-surge-worldwide.

[14] Greger, *How to Survive a Pandemic*, 110.

[15] World Health Organization, https://www.who.int/en/news-room/detail/14-12-2017-up-to-650-000-people-die-of-respiratory-diseases-linked-to-seasonal-flu-each-year. The exception, of course, is the Spanish flu epidemic, which killed upward of 50 million. More on this later.

[16] David A. Koplow, *Smallpox: The Fight to Eradicate a Global Scourge* (Berkeley: University of California Press, 2003).

[17] Greger, *How to Survive a Pandemic*, 110.

[18] Suzanne Austin Alchon, *A Pest in the Land: New World Epidemics in a Global Perspective* (Albuquerque: University of New Mexico Press, 2003.

[19] T. McMichael, *Human Frontiers, Environments, and Disease* (Cambridge: Cambridge University Press, 2001).

[20] Wolfe, Dunavan, and Diamond, "Origins of Major Human Infectious Diseases."

[21] McMichael, *Human Frontiers, Environments, and Disease.*

[22] M. J. Rodrigo and J. Dopazo, "Evolutionary Analysis of the Picornavirus Family," *Journal of Molecular Evolution* 40 (1995): 362–371 as cited in E. F. Torrey and R. H. Yolken, *Beasts of the Earth: Animals, Humans, and Disease* (New Brunswick, NJ: Rutgers University Press, 2005), 50.

With tropical diseases, the animal reservoir tends to consist in wild animals rather than domesticated ones. The HIV virus that results in AIDS originated in primates: it developed out of simian immunodeficiency virus (SIV), which resides harmlessly in apes.[23] Overall, globally, the virus has been estimated to have infected over 60 million people, resulting in approximately 32 million deaths.[24] In many cases of tropical diseases, while the origin of the disease is an animal reservoir, the vector of transmission is an insect of some form. Chagas disease is found in a wide range of wild and domestic mammals, but it is typically spread by the *Triatominae* or "kissing bug." The original animal reservoir of dengue fever is believed to be Old World primates. But it is typically transmitted by the *Aedes* mosquito. The original animal reservoir of *Falciparum* malaria is thought to be wild birds, perhaps now extinct. For *Vivax* malaria, however, the original reservoir is thought to be Asian macaque monkeys, although this reservoir has now expanded to occasionally include New World monkeys also.[25] Both, however, are largely transmitted through female Anopheles mosquitos. Globally, malaria kills hundreds of thousands of people annually, mostly in Africa.[26] Wild and domestic ruminants form the animal reservoir of both East and West African sleeping sickness.[27] The vector of transmission of the *Trypanosoma brucei* parasite that causes this disease is the tsetse fly. The original animal reservoir for Hepatitis B was apes,[28] although since this has now mutated into an exclusively human virus, it no longer has an animal reservoir. Yellow fever also originated in apes.[29]

Temperate and tropical diseases differ with respect to the specific pathology trajectories. As a general rule, with various exceptions on both sides, temperate diseases tend to be acute. You will get them, you will either die or live, and, if the latter, the disease will bid you adieu, usually within a few weeks at most. Tropical diseases, however, often tend to be chronic ones. AIDS will last a lifetime—however long that may be. Chagas disease lasts from months to decades. East African sleeping sickness can last for up to 9 months, after

---

[23] Wolfe, Dunavan, and Diamond, "Origins of Major Human Infectious Diseases."
[24] https://www.who.int/gho/hiv/en/.
[25] Wolfe, Dunavan, and Diamond, "Origins of Major Human Infectious Diseases."
[26] According to the WHO, malaria was responsible for 435,000 deaths in 2017, 403,000 of which occurred in the WHO African region. See https://www.who.int/gho/malaria/epidemic/deaths/en/.
[27] Wolfe, Dunavan, and Diamond, "Origins of Major Human Infectious Diseases."
[28] Wolfe, Dunavan, and Diamond, "Origins of Major Human Infectious Diseases."
[29] Wolfe, Dunavan, and Diamond, "Origins of Major Human Infectious Diseases."

which time, without treatment, you will be dead. West African sleeping sickness can last for years or even decades without treatment.

The Making of a Zoonotic Pathogen

COVID-19 probably began life as an unassuming coronavirus, SARS-CoV-2, harbored in horseshoe bats, to which it apparently did little harm—like most viruses harbored in bats, in fact. From there it, somehow, became the scourge of humanity in the twenty-first century. The pangolin was an initial favorite to have the played the role of intermediate host—a stepping stone between bat and human. Then it fell out of favor for a while. But now, alas for the poor pangolin, to which we humans have done so much harm, it now appears to be back in contention again.[30] In coming to afflict humans in this dramatic way, SARS-CoV-2 is, mercifully, unusual. Most animal pathogens do not infect humans at all: they remain, resolutely, animal pathogens for all their days. For an animal pathogen to infect a human being, various things have to happen first. Wolfe, Dunavan, and Diamond identify five stages in the transformation of an animal pathogen into one that exclusively infects humans.[31]

> *Stage 1.* The pathogen exists in animals but is not transmitted to humans under *normal* conditions. We humans are, of course, good at inventing abnormal conditions—technologies such as blood transfusions, organ transplants, hypodermic needles, and so on, which can sometimes transfer to us pathogens we otherwise would not have acquired. For example, many malarial plasmodia are specific to their usual host species and would not infect humans, except in these technologically facilitated circumstances.
>
> *Stage 2.* A pathogen exists in animals and has been transmitted to humans under natural conditions. This transmission is known

---

[30] The horseshoe bat/pangolin hypothesis reflects, at the time of writing, the latest thinking on its origin. By the time this book comes out, of course, all this may have changed. This is based on comparative genetic analysis carried out by S. Lau, K. Hayes, A. Luk, C. Wong, K. Li, L. Zhu, Z. He, J. Fung, T. Chan, S. Fung, and P. Woo, "Possible Bat Origin of Severe Acute Respiratory Syndrome Coronavirus 2," *Emerging Infectious Diseases* 26, no. 7 (July 2020): 1542–1547. https://wwwnc.cdc.gov/eid/article/26/7/20-0092_article.

[31] Wolfe, Dunavan, and Diamond, "Origins of Major Human Infectious Diseases."

as *primary transmission* or *primary infection*. However, with a stage 2 pathogen, there is no *secondary transmission* or infection. That is, the pathogen does not become transmissible between humans, or becomes transmissible only in the most unusual of circumstances. Rabies is a good example of a stage 2 pathogen. It can be passed from a rabid animal to a human through saliva, but it is not, in general, transmitted between humans. This is because of differences in the behavior of infected humans compared to infected animals. Infected dogs will often bite. But infected humans rarely bite. Therefore, transmission between humans is possible but unusual. Other examples of stage 2 pathogens include anthrax, tularemia bacilli, Nipah, and West Nile viruses.[32]

Stage 3. A stage 3 zoonotic pathogen is one that is transmissible between humans, but where the efficiency of transmission is rather low. The result is that while there are outbreaks of diseases caused by stage 3 pathogens, these are generally local rather than pandemic or epidemic, and tend to die down quite quickly. This is good news, because stage 3 pathogens can be very nasty. Ebola and Marburg viruses—both causing hemorrhagic fever—are well known. The *monkeypox* virus is also a stage 3 pathogen.

Stage 4. This is the most complex transformational stage that an animal pathogen might go through and Wolfe, Dunavan, and Diamond usefully break it down into three distinct stages. The general idea is that a stage 4 pathogen will have a natural (or *sylvatic*) cycle of infecting humans through primary infection from an animal host. However, subsequently, there will be long sequences of human-to-human-to-human transmissions, or *secondary infections*, with no further contribution from the animal host. This general idea can be divided into three substages.

Stage 4a. Human-to-human transmission occurs, but it is nowhere near as important as the sylvatic cycle in driving the spread of the disease. For example, the spread of Chagas disease between humans depends crucially on the seasonal life cycle of the *Triatominae* insect, commonly known as the *kissing bug*. Similarly, with the

---

[32] Examples cited by Wolfe, Dunavan, and Diamond, "Origins of Major Human Infectious Diseases."

yellow fever virus, the most significant factor driving human infections will be the actions of the *Aedes* or *Haemagogus* mosquitoes, which carry the disease from infected monkeys to humans, first in the jungle and then in urban settings.

Stage 4b. With a pathogen at this stage of development, both the sylvatic cycle and secondary, human-to-human transmission are equally important in driving the spread of the disease. Examples include dengue fever, especially in forested regions of West Africa or Southeast Asia.

Stage 4c. In this stage of the pathogen's development, the greatest form of transmission is human to human, although the animal reservoir, or primary infector, can still play a nonnegligible role. Examples of this are influenza A, cholera, typhus, West African sleeping sickness, and also, topically, SARS-CoV-1and SARS-CoV-2. You might certainly catch COVID-19 from the pangolin you have been planning to eat—if the pangolin is, indeed, the intermediate infector. But, assuming the number of people you meet is far greater than the number of pangolins you eat, then you are more likely to catch it from people than from pangolins.

Stage 5. A stage 5 pathogen is one that has evolved to become exclusive to humans. Now, strictly, it is now no longer a zoonosis but a pathogen with a zoonotic origin. Examples of exclusively human pathogens with zoonotic origins include *falciparum* malaria, measles, mumps, rubella, smallpox, and syphilis.

Most of the diseases that afflict humans, including some of the nastiest, are either zoonoses or diseases of zoonotic origin. It is rather important, then, to identify exactly why we keep getting them. To do this, we need to look at why pathogens go through these five transformational stages. Stage 1 pathogens need not worry us, the odd blood transfusion or organ transplant aside. Stage 2 pathogens may seem only a little more disconcerting. Rabies, for example, is endemic in the United States, and occasionally there are little outbreaks in various localities. But it is not the sort of thing over which one loses sleep. Nevertheless, stage 2 is also a watershed—a proof of concept, if you will. The pathogen has demonstrated the ability to infect the new host cells. From there it merely has to work out how to get transmission going between its new hosts organisms. By the time it gets to stage 3, and that all-important stage 4, we tend to begin taking them seriously. But it is important to remember that the vast majority of animal pathogens never get beyond stage 1. The crucial question we should ask ourselves, then, is that, for those pathogens that venture beyond stage 1, especially if they get to stages 3 and 4: Why does this happen? Why do some pathogens follow this trajectory while most stay firmly rooted in stage 1?

How to Make Friends and Infect People

The short answer is: *dumb luck*, or the lack thereof, depending on your perspective. A marginally longer answer that means essentially the same thing as dumb luck is: *random mutation reinforced by natural selection*. When a novel pathogen finds itself in a new host, it needs to bind, in some way, to the host's cells. This can prove difficult as the novel host's cells may be quite different from the ones to which it has become accustomed. However, the pathogen is always churning out copies of itself, and there may be small variations between these copies, the result of little imperfections in the copying process. Occasionally, just occasionally, an imperfection in the fidelity of the copying process can confer an advantage on the copy, making it more able to lock on to and infect the cells of a new host. Admittedly, the chances of this happening on any given occasion—where the new pathogen copy and host cell are aligned—are very minuscule: not quite the army of monkeys randomly typing out the complete works of Shakespeare kind of minuscule, but slim enough. But suppose the pathogen gets many bites of the cherry. That is, the pathogen has millions upon millions of opportunities to align itself with the cells of the prospective new host. Then, its chances of success rapidly improve. Eventually, if the pathogen gets enough opportunities to try out the random mutation maneuver, it may very well strike it lucky.

With respect to potentially zoonotic pathogens, a mutation can be more or less *relevant*. A pathogen can mutate all it likes, but if there is no potential host in the vicinity, all this mutation will be in vain. A relevant mutation, therefore, is one that occurs in the presence of a potential host. There are, broadly speaking, two factors that will determine a pathogen's chances of successfully adapting to a new host. The first of these is how many relevant mutations it is *allowed* to make. The second is how many relevant mutations it is *required* to make.[33] Let's, first, look at the number of mutations (per unit time) a given pathogen will be allowed to make. This will be a function of several factors.

The first factor is the *abundance of the existing or original host*. From the point of view of your aspiration to jump to a new host, the more there is of the existing host, the better. If, for example, you are a virus specializing exclusively in the Madagascar pochard duck—reputedly the rarest bird in the world—then the likelihood of you getting transmitted to a new potential host, will, generally speaking, be rather low. Sadly, no matter how much you might be suited to the new potential host—no matter how easy it would be for you to adapt and how much damage you could wreak once you had—you are unlikely to get a chance to show what you can do.

The second factor is the *proportion of the existing animal host population that is infected*. The abundance of the host animal is one thing, and the fraction of the host population is quite another. If, for example, you are a virus that specializes in mallards, but have, for some reason or another, only managed to infect a tiny handful of the roughly 11 million mallards that inhabit or visit the United States every year,[34] then your opportunities for encounters with other potential hosts are also likely to be relatively restricted.

The third factor is the *frequency of encounters between the animal host and potential human host*. For an aspiring pathogen, looking for ways to jump the species barrier, encounters are everything. You may have managed to infect a large percentage of the host animal species, a host that comprises not inconsiderable numbers, but if you are prevented from ever encountering other species, this will count for nothing. Generally speaking, the possibility of encounters depends on *encroachment* on territory—in one direction or another. Either a new potential host encroaches on your territory in some way, or you encroach on theirs, or both.

The fourth, and final factor, perhaps stretches the notion of being allowed to make mutations—for here, what mutations you are allowed to make depends on your own biological history. But it's near enough. I am talking of the *pathogen's genetic flexibility*. Large numbers of original host, large percentage of original host infected, and frequent encounters with other prospective host species will count for nothing if you are unable to mutate, or

---

[33] As we shall see in the next chapter, zoonotic pathogens around today have been born into a truly serendipitous epoch, where various circumstances are arranged like—no pun intended—ducks in a row to give them the maximum chance of success.

[34] https://www.ducks.org/conservation/waterfowl-surveys/2018/duck-numbers.

you do so very slowly. You are going to have to make some adjustments in order to align yourself with the new host environment in which you find yourself. If you are one of those sedate, or overly inflexible, pathogens that just aren't able to adjust quickly enough, then you may not make it. For a pathogen with species-hopping aspirations: you snooze, you lose.

These four factors—abundance of original host, percentage of original host infected, frequency of encounters, and biological flexibility—pertain to the number of mutations the pathogen is going to *allowed* to make in its quest to jump species. When we switch to the number of mutations it is going to be *required* to make, the picture is less cluttered. Here, one factor is of over-riding importance: the *phylogenetic distance between* the pathogen and its prospective new host organism. The greater this phylogenetic distance, the more mutations the pathogen will be required to make to acclimatize to its new host—and, therefore, the less likely it will be able to do so. For an avian influenza virus, specializing in ducks, then a sagacious next move would be to jump to chickens rather than chimpanzees, geese rather gorillas. Baby steps. You'll get to primates in the end, no doubt, but you need to take it, as the old cliché goes, one game at a time. The best way to get to primates is by using a series of *intermediate hosts*.

Let's look at some examples of these general principles at work. Consider, first, *chimpanzees* (and other great apes). Counting against pathogens leaping from them to us is the relatively low abundance of chimpanzees, combined with relative infrequency of encounters between us and them. What counts in favor of pathogen transmission is the tiny phylogenetic distance between us and them. The result of this combination of factors has been a succession of nasty diseases, including HIV/AIDS, Hepatitis B, Ebola, and other hem-orrhagic fevers, such as Marburg, Lassa and Rift Valley fever, and probably yellow fever.

Take the Ebola virus, for example. The original animal source of this virus is the fruit bat. However, the fruit bat regularly transmits the virus to other, intermediate hosts, including antelopes and porcupines and also monkeys, chimpanzees, and gorillas. The phylogenetic distance between us and chimpanzees is minuscule, meaning that any variant of the Ebola virus found in these animals is already highly adapted for human infection. As a result, when humans catch Ebola, it is usually through chimpanzees and gorillas. What, one might ask, about the relative infrequency of encounters between us and them? The answer is that we have ways of undermining that—one of which is the commercial bushmeat industry. Much of this industry is driven by commercial logging and mining operations in the African rainforest.[35] Where commercial extractive companies go, vendors—selling food and other concessions—quickly follow. Supplying workers with bushmeat—wild animals killed for human consumption—is much cheaper than transporting in domesticated meat. Therefore, these kinds of companies encourage it, and bring in an army of commercial bushmeat hunters, to supply the workers. These bushmeat victuals include more than 26 species of primates, endangered chimpanzees and gorillas among them. This is one of the major causes of recent outbreaks of Ebola.[36]

Rats, of course, are notorious spreaders of disease. Certain factors count against their being able to infect humans—the greater phylogenetic distance between us and them, in comparison with the great apes at least. More than offsetting this, however, is their high abundance, the typically high percentage of infected population (a function of their population density), and the relatively high frequency of encounters between us and them. The results are well known: various forms of plague—bubonic, pneumonic, septicemic— Hanta virus, Tularemia, Lymphocytic Choriomeningitis, and so on.

The greater phylogenetic distance between us and *domestic livestock*— including cattle, pigs, sheep, goats, and domestic fowl—counts against their ability to successfully transmit pathogens to us. However, this impediment is more than offset by the high incidence of encounters between us and them— we breed them, feed them, milk them, slaughter them, and so on—all part of the practice of raising them for food. The result of this convergence of factors has been more than half of all temperate diseases and more than a few tropical diseases. The category of domestic livestock is, of course, a broad one, and degrees of genetic distance vary. Chickens are, obviously, more genetically distant from us than pigs. However, the fact that different families of domestic livestock are often raised in close proximity means that the use of intermediate hosts in the transmission of pathogens is commonplace. As we shall see in the next chapter, pigs are a readily available intermediate—or bridging—host between us and domestic fowl.

Elephants are a useful test case. There is a relatively large phylogenetic distance between us and them—in comparison with the great apes, for example. They exist in low abundance. And the frequency of encounters between us

---

[35] For discussion, see Greger, *How to Survive a Pandemic*, 115–117.
[36] W. Karesh, R. Cook, E. Bennett, and J. Newcomb, "Wildlife Trade and Global Disease Emergence," *Emerging Infectious Diseases* 1, no. 7 (2005), 1000–1002, cdc.gov/ncidod/EID/vol11no07/05-0194.htm.

and them is very low. The result is exactly what we should expect. There have been no known diseases transmitted from elephants to humans.

Bats are another usefully illustrative case. On the one hand, they are highly abundant, and also most bats seem to carry astonishingly high levels of infectious pathogens (which actually don't seem to unduly inconvenience them—we shall look at possible reasons for this in the next chapter). On the other hand, counting against this abundance of bats and their pathogens is the relatively large phylogenetic distance between us and them, and also the relative infrequency of direct bat-human encounters. The results are well documented: rabies, Nipah virus, SARS-CoV-1, and SARS-CoV-2, among many, many others. The reason for this is bat's prodigal use of intermediate hosts. As we saw in the case of Ebola, most humans acquire this virus from infected apes and monkeys rather than from bats. Bats rarely directly infect humans with rabies. Typically, an infected dog—an intermediate host that will itself have picked up the virus from another intermediate host such as a fox or coyote—is the direct vector of transmission of rabies to humans. SARS-CoV-1 found its way from bats to humans via the civet cat, which is a delicacy in some parts of the world. SARS-CoV-2, some current thinking suggests, bridged the gap between horseshoe bats and humans by means of the pangolin, the world's most trafficked animal, prized for its flesh and the alleged medicinal qualities of its (powdered) scales.

Please Allow Me to Introduce You

As the aforementioned cases illustrate, the use of intermediate hosts is a widespread feature of interspecies pathogen transmission. Often, however, it is not simply that we find an intermediate host or hosts present in the transmission chain. Rather, it is that we, through our own actions, put that intermediate host there. If I get bitten by a rabid coyote, that is one thing. We might ask questions about how the coyote and I came to be sharing the same piece of land. But we can imagine a case where neither of us was systematically encroaching on the home of the other. I was out for a largely noninvasive walk in the countryside, and so on. That, as I say, is one thing. Sometimes, shit just happens. But the case of Ebola, and the two versions of SARS, are quite different. In these cases, the intermediate hosts are there because, and only because, we put them there. Chimpanzees and gorillas would not be intermediate hosts of Ebola if we hadn't insisted on hunting and killing them.

They would have Ebola, admittedly, but they wouldn't be intermediaries in the transmission chain from fruit bats to humans. The civet cat would not have been an intermediate host in the SARS CoV-1 transmission chain if we did not insist on eating it. The same is true of the pangolin's role in the SARS-CoV-2 transmission chain: it is only there because we put it there. Moreover, in some cases, it is not simply that we put in place an intermediate host in a pathogen transmission chain that ultimately leads to us. Sometimes we do more than even this. Sometimes *we actively orchestrate an introduction between original and intermediate host*. We both put in place the link and enthusiastically engineer the process of linking.

The Nipah virus provides a very good example of this, and it foreshadows some of the themes to be developed in the next chapter. The Nipah virus is named after the village, Sungai Nipah, in which its first human fatality occurred. In 1997, some 10 million acres of tropical rainforest in Sumatra and Borneo found themselves on fire, in the service of the then nascent palm oil industry. This, quite predictably, resulted in a mass migration of fruit bats. Some of these ended up in mango trees that were in close proximity to some rather massive Malaysian pig farms. The bats dribbled mango fruit and sprayed urine into the pig pens. Sadly, the saliva and urine both contained a virus. The pigs became sick first, developing a ghastly cough, followed by internal hemorrhaging. Neurological symptoms—spasms and paralysis—were quickly followed by death. Humans became sick soon after and displayed similar symptoms: high fever, headaches, convulsions, and swelling of the brain, and then death. The outbreak lasted 7 months, spreading all over the Malaysian peninsula and even reaching northern Australia. The human case mortality rate was around 40%, and some survivors would experience relapsing brain infections for several months afterward.

The genesis of the Nipah virus in humans and pigs combines several important elements. First, there is the environmental disturbance—the destruction of the rainforest—engineered by us. In engineering this, we managed to introduce the animal reservoir of a virus to another species, pigs, which also provided a suitable environment for the virus. The pigs then acted as intermediate hosts, passing the virus on to us. The mere having or carrying of a virus is not enough for the pigs to qualify as intermediate hosts. For this, we needed to do something else. The pigs became intermediate hosts only because of what we did to them—farmed them and ate them. In short, we engineered the introduction that made the pigs into hosts—in the sense of carriers—of the virus. And we engineered the circumstances in which these hosts could become specifically *intermediate* hosts—links in the transmission chain that led from fruit bats to us.

Aggressively Symbiotic Bad News

A pathogen can live, often quite innocuously, in an animal reservoir species for a long, long time. However, when it jumps to a new host, it is quite common for the pathogen to become notably more, let us say, *weaponized*. We saw this in the case of the Nipah virus, for example. In bats, it seems to cause no problems. But when it jumps from bats to pigs, and from pigs to humans, it becomes a genuinely nasty piece of work. Bats are a notorious reservoir of many viruses—including the coronaviruses and retroviruses that ultimately became SARS-C0V-1 and 2, and HIV, respectively. But these viruses do not, in general, seem to trouble bats. It is only when they jump to a new species that problems ensue. Similarly, all human flu viruses ultimately began life as a harmless virus that inhabited the intestines of ducks. In the jump to another species, the influenza becomes potentially lethal. This is a widely recognized feature of zoonotic pathogens. In a similar vein, the H5N1 avian influenza is seemingly harmless in ducks, but when it jumps to chicken becomes very nasty and highly lethal.

The reasons for this increased virulence subsequent to a jump of species are not well understood. Part of the story—certainly not all—might involve the phenomenon of *aggressive symbiosis*. The underlying idea is that certain viruses can confer a selective advantage upon creatures in which they have become endemic and settled. For example, a form of herpes virus, *Herpesvirus saimiri*, is endemic in many populations of squirrel monkeys. If another species, in this case the marmoset, encroaches on the squirrel monkey's territory, they acquire the virus, which, in them, takes lethal form. It is relatively easy to work out the evolutionary story of how this might have happened. Once again, random mutation lies at the heart of this story. Suppose a virus mutated to have these properties: harmless to squirrel monkeys, lethal to marmosets. Squirrel monkeys and marmosets are competitors. Therefore, groups of squirrel monkeys that had this virus would have an advantage over other groups that did not. Thus, the infected groups would have thrived at the expense of their noninfected peers, thereby out-reproducing them. In this way, the virus would have spread throughout the squirrel monkey population.

The idea of aggressive symbiosis is not uncontroversial and presumably of limited scope. With respect to pathogenic virulence subsequent to a species jump, if it provides an explanation at all, it will only be for species that are genetically and environmentally close enough to count as competitors for resources. But what is uncontroversial is the idea that when a pathogen jumps from one species to another it often transforms from unassuming or harmless in the original host to positively lethal in the new one. The current pandemic circumstances in which we find ourselves, and the worrying *probability* of far worse pestilences to come, should provide pause for thought. Through our actions, we engineer the introduction of animals to each other, significantly increasing the probability of a pathogen jumping the species barrier and transitioning to lethal. And through our actions, also, we significantly raise the probability that this newly lethal pathogen will, eventually, turn back around and bite us. At the heart of this destructive, reckless, behavior is an old habit than we could never quite shake: eating animals. The specific role of this habit in our current, and almost certainly future, plague circumstances is the subject of the next chapter.

11

One Hundred Years of Ineptitude

No Flying Cars, Then?

It's funny how life turns out. When I was a boy in 1970, I thought that we would all be learning to drive flying cars by now. Instead, we're all learning how to wash our hands properly again. There is a well-known saying, of not entirely clear provenance, but usually attributed to the philosopher George Santayana that "Those who cannot remember the past are condemned to repeat it." COVID-19 wasn't inevitable, but it stems from a failure to remember the past—recent and not so recent. And while governments around the world were seemingly taken by surprise, it shouldn't really have been that surprising. The original animal reservoir of the SARS-CoV-2 virus, the virus that causes the disease COVID-19, was probably a bat—the horseshoe bat appears to be a leading contender.[1] The virus, the evidence also suggests, probably did not jump from the bat directly to us but proceeded via an intermediate host. The pangolin was an early bookies favorite to be this host, before fading somewhat in the ensuing months. But I gather it has been making a determined run down the rails in recent weeks and may indeed turn out to be the intermediate host.[2] If this is correct, there is one and only one reason why the world finds itself in the grip of a pandemic. We eat bats, and we eat pangolins.[3] Or is that two reasons? Either way, without these dietary predilections there would almost certainly have been no COVID-19 pandemic. In this, COVID-19 joins a long list of viral predecessors.

This kind of route—bat to intermediate host to us—is expected because we have been here before. Before COVID-19, there was what was known then simply as SARS (severe acute respiratory virus), caused by the SARS-CoV-1 virus as it has now become known. This coronavirus—named for the

---

[1] See S. Lau, H. Luk, A. Wong, K. Li, L. Zhu, Z. He, and P. Woo, "Possible Bat Origin of Severe Acute Respiratory Syndrome Coronavirus 2," *Emerging Infectious Diseases* 26, no. 7 (2020): 1542–1547. https://dx.doi.org/10.3201/eid2607.200092. https://wwwnc.cdc.gov/eid/article/26/7/20-0092_article.

[2] Lau et al., "Possible Bat Origin of Severe Acute Respiratory Syndrome Coronavirus 2."

[3] We also use them, rather unsuccessfully I gather, for medicinal purposes.

World on Fire. Mark Rowlands, Oxford University Press. © Oxford University Press 2021.
DOI: 10.1093/oso/9780197541890.003.0011

crown of spikes emanating from the virus body—probably also originated in horseshoe bats[4] and used the civet cat as an intermediate host-bridge to humans. Civet cats provided an appropriate intermediate host because, and only because, humans eat them. Among other things, the penis of the civet cat is highly regarded for its (alleged) aphrodisiac properties. SARS-CoV-1, in humans, emerged in the Guangdong province in southern China in November 2002, and by the end of 2003 it had infected an estimated 8096 people worldwide, resulting in 774 deaths. While this might seem pretty minor on the epidemiological scale of things, its significance is not restricted to its acting as a harbinger of its more famous cousin. Its case mortality rate of nearly 10% is also a little troubling, because it demonstrates the kind of virulence of which coronaviruses are capable.

You might think of SARS-CoV-1 and SARS-CoV-2 as, together, providing us with a proof of concept. For quite some time, coronaviruses had been dismissed as minor irritants—causes of nothing more significant than the common cold. With SARS-CoV-1, there is proof that a coronavirus could produce a disease with high mortality rate.[5] In recent months, the world has been grappling with a virus that has a mortality rate of, current evidence seems to suggest, somewhere between 0.5% and 1% in the population as a whole.[6] This is five to ten times as deadly as the average seasonal flu, but far short of the 10% estimated for SARS-CoV-1. This grappling has seen much of the world shut down, and national economies devastated, perhaps for years or decades to come. Imagine what things would be like if its mortality rate were near 10%. The saving grace of SARS-CoV-1 was that its level of infectivity—its R_0 number—was rather low. But SARS-CoV-2 shows that it is possible for a coronavirus to be highly infectious. Prior to masks, social distancing, and other measures, the R_0 of SARS-CoV-1 was estimated to be between the 1.4 and 6.0 mark—making it not as infectious as measles but significantly more infectious than the seasonal flu.[7] It may be that when

[4] A. Banerjee, K. Kulcsar, V. Misra, M. Frieman, and K. Mossman, "Bats and Coronaviruses," *Viruses* 11, no. 1 (2019), 41, doi:10.3390/v11010041 https://www.ncbi.nlm.nih.gov/pmc/articles/PMC6356540/.

[5] A fact later confirmed by the even more lethal MERS-CoV virus. *Vide infra.*

[6] See, for example, S. Mallapaty, "How Deadly Is the Coronavirus? Scientists Are Close to an Answer," *Nature* 16 (June 2020), https://www.nature.com/articles/d41586-020-01738-2. As I put the finishing touches to this book (August 5, 2020), the World Health Organization (WHO) seems to favor an estimate of 0.6. Or 1 in every 167 people.

[7] At least, in an as yet non-peer-reviewed preprint. K. Linka et al., "The Reproduction Number of COVID-19 and Its Correlation with Public Health Interventions," *medRxiv* preprint, 2020, http://medrxiv.org/cgi/content/short/2020.05.01.20088047.

SARS-CoV-3 debuts, it combines the worst of both of its predecessors—high mortality coupled with high infectivity. If so, things will obviously get bad very, very quickly. If we didn't eat bats, civets, and pangolins, it is likely that neither SARS-CoV-1 nor SARS-CoV-2 would ever have become a disease that infected humans, and certainly not a pandemic disease.

We establish the links required for viruses such as SARS-CoV-1 and SARS-CoV-2 to jump from bats to us. We introduce bats to civets and pangolins, and we introduce civets and pangolins to ourselves—specifically, to our stomachs. Typically, these introductions take place in particularly torrid circumstances. In a wet market, such as the Wuhan Sea Food market, theorized to be ground zero for the current COVID-19 epidemic, animals of different species—including bats and pangolins, but also, apparently, wild rabbits, peacocks, deer, snakes, wolf cubs, and many others—are crammed into crates and cages, often stacked on top of each other. Prima facie, a situation involving hordes of highly stressed animals crammed together in close vicinity—urinating, defecating, and eventually brutally slaughtered— does seem to be something close to an ideal petri dish for the mixing of viruses such like SARS-CoV-1 and SARS-CoV-2. These are the sorts of circumstances, in fact, in which a horseshoe bat might very well be able to infect a pangolin with a variant of the SARS virus.

When a zoonotic pathogen takes hold in a human population, it is almost always because we have engineered the circumstances for this happen. Specifically, we engineer introductions between animals of different species, including ourselves. And we do this because we want to eat one or more of the animals introduced. A brief perusal of all the major disease outbreaks in the last century or so highlights the role humans have played in engineering opportunities for pathogens to mingle, to get to know each other, and ultimately to get to know us. The West African Ebola outbreak of 2014 was the most widespread outbreak of the Ebola virus in recorded history. Beginning in Guinea, it quickly spread to Liberia and Sierra Leone, causing widespread socioeconomic disruption, and resulting in an estimated 11,323 deaths. Ebola is a viral hemorrhagic fever that causes severe internal bleeding, organ failure, and, in the 2014 outbreak, death in over 50% of cases. Ebola is also a zoonosis, originating in fruit bats, with chimpanzees and gorillas, among other animals, also susceptible to the virus and almost certainly acting as intermediate hosts.

As I mentioned in the previous chapter, outbreaks of this virus are typically caused by one of two things. The first is that we eat infected animals—bushmeat, smoked chimpanzee and gorilla meat, and the flesh of various other primates and monkeys. In doing so, we make these animals into intermediate hosts of the virus. Having a virus, or other pathogen, that you have acquired from something else is one thing. Being an intermediate host of that virus is quite another. You only become that when you, in turn, pass the virus on to something else. That a gorilla, chimp, or other primate has the Ebola virus does not make it an intermediate host. It becomes an intermediate host only when we decide to eat it. The second cause of Ebola outbreaks—and we saw the role this factor also played in the outbreak of Nipah—is that we encroach upon the ancestral home of the original host of the virus.[8] Sometimes we do both at the same time. Through logging we disturb the fruit bat's home, and to feed the loggers we employ commercial bushmeat operators.[9]

The *Middle East respiratory syndrome* (MERS) outbreak of 2012 was hardly a major one. Largely confined to the Arabian Peninsula, with a smaller outbreak in South Korea, this was nevertheless worrying due to its high mortality rate of around 35%. This was again the result of a coronavirus, MERS-CoV, carried by bats—in this case, it seems, Egyptian tomb bats, lesser mouse-tailed bats, and Kuhl's pipistrelle. While bats were the original host, the intermediate host of MERS seems to be the dromedary camel—a delicacy in the Middle East. As of November 2019, there were 2494 confirmed cases of MERS, resulting in 858 deaths.

Since the onset of the HIV/AIDS epidemic, in 1981, it is estimated that roughly 75 million people have been infected with the HIV virus, resulting in the deaths of around 32 million.[10] Here, the animal reservoir consisted in chimpanzees and gorillas that carried *simian immunodeficiency virus* (SIV). This was harmless in chimpanzees and gorillas, but when transmitted to humans, it mutated into the deadly HIV. This transmission almost certainly occurred via the consumption of infected bushmeat. The epidemic was brought to us by the habit of eating animals. If we didn't eat bushmeat, HIV/AIDS probably would have never happened.

As I suggested earlier, sometimes, in the face of our own culpability, great solace can be found in asking the wrong questions or jumping to the wrong conclusions. The coronavirus pandemic has seen the finger of blame

---

[8] W. Na, N. Park, M. Yeom, and D. Song, "Ebola Outbreak in Western Africa 2014: What Is Going on with Ebola Virus?" *Clinical and Experimental Vaccine Research* 4, no. 1 (2015): 17–22. doi:10.7774/cevr.2015.4.1.17 https://www.ncbi.nlm.nih.gov/pmc/articles/PMC4313106/.

[9] See Michael Greger, *How to Survive a Pandemic* (New York: Flatiron Books, 2020).

[10] World Health Organization, Global Health Observatory (GHO) Data, https://www.who.int/gho/hiv/en/.

pointed squarely at bats. Bats, that is, and the Chinese people for their, as we see it, misbegotten habit of eating wild animals. In reality, however, it is not eating specifically *wild* animals that is the problem. The real problem is eating animals—period. To see this, we merely need to realize that there are some outbreaks I have not yet included on the earlier list. The earlier noted outbreaks are noninfluenza outbreaks—coronaviruses, a filovirus, and a retrovirus. For many researchers working in the area of emerging infectious diseases, I think it is fair to say that it is influenza, more than any other type of virus, that keeps them awake at night. And in the case of influenza, the road to infection is populated by domestic animals at least as much as their wild counterparts.

The Spanish Flu

Imagine. There is a disease in circulation. The first signs that you have this disease are fever and muscles aches. But in a matter of days, you may well find yourself bleeding from your nostrils, ears, and eye sockets. Parts of you, extremities—fingers, toes, penises, the usual suspects—will turn black. The miasma of bodily decay will surround you even before you are dead. Your body will turn a dark blue, a symptom of a deficit of oxygen with which your lungs can no longer supply you. A final spasm, blood erupts from your mouth. You die drowning, your lungs full of your own bloody secretions.[11] Pretty nasty. Now imagine that one-third of the world's population will get this disease. And, of those, around 10% will die. What you have imagined, in effect, is the Spanish flu of 1918–1920.[12]

The influenza virus comes in various forms defined in terms of spikes, made of hemagglutinin (H) and neuraminidase (N), that they use to lock onto and infect host cells. There are 18 types of hemagglutinin spike (H1 through H18) and 11 different types of neuraminidase spike (N1 through N11). The Spanish flu is classified as H1N1, signifying that it comprises the first type of hemagglutinin spike and the first type of neuraminidase spike. This 1918–1920 iteration of the H1N1 virus was the most serious human health event in modern times. Globally, an estimated 500 million people were infected—roughly one-third of the global population at that time. At

[11] Greger, *How to Survive a Pandemic*, 25–30.
[12] Some date this flu as 1918–1919. There was, in fact, a fourth wave in 1920, but it was relatively minor compared to the first three.

least 50 million people died,[13] and the real figure may be closer to 100 million.[14] The Black Death and other plagues of the Middle Ages killed more people, but they took centuries and several iterations to do so. As far as we know, never before in the field of human disease had so many died in so short a time span.

The Spanish flu was unusual in another respect: mortality was highest in young and otherwise healthy groups. In fact, mortality peaked in the 20–34 age group. This is in stark contrast to the seasonal flu, which tends to take the oldest and youngest members of society.[15] Nor were the baleful effects of the flu limited to those who died, and the impacts of the pandemic also extended long past the 1918–1920 window commonly ascribed to the four separate waves of the disease. In the subsequent decade, roughly a million people were afflicted with a form of encephalitis, which some researchers believe was a consequence of neurological damage inflicted by the virus. Babies, in utero in affected areas, had shortened lifespans and sometimes lifelong disability.[16]

The Spanish flu was misnamed. It didn't begin in Spain. The First World War had seen a moratorium on accurate reporting of the news in the countries involved. If you are trying to both conduct a war and grapple with a flu virus, the last thing you want to do is let your enemy know about the latter—even though they are probably grappling with it, too. Spain, however, was not involved in the war, and so was under no such reporting restrictions. As a result, most of the early reported news of this disease was to come from Spain. But the virus didn't originate there, and it is not entirely clear where it did originate: evidence can be found in favor of the United States, France, the United Kingdom, and China, but this evidence is far from definitive.

While its ultimate geographical origin may be unclear, there is no similar doubt about its biological origin. That origin was, ultimately, *avian*. There is, actually, a rather gripping tale of the discovery of this avian origin of the Spanish flu virus. The demonstration of this origin was provided by Taubenberger and colleagues, who managed to reconstruct the virus's

[13] CDC, https://www.cdc.gov/flu/pandemic-resources/1918-pandemic-h1n1.html.
[14] N. Johnson and J. Mueller, "Updating the Accounts: Global Mortality of the 1918–1920 'Spanish' Influenza Pandemic," *Bulletin of the History of Medicine* 76 (2002): 105–115.
[15] Greger, *How to Survive a Pandemic*, 27.
[16] D. Almond, "Is the 1918 Influenza Pandemic Over? Long-Term Effects of In Utero Influenza Exposure in the Post-1940 U.S. Population," *Journal of Political Economy* 114, no. 4 (2006): 672–712.

genome.[17] To do this, Taubenberger required the (preserved) body of a long-dead victim of the flu. The gripping tale in question concerns how he was provided with this body—or, at least, relevant parts of it—a tale involving multiple trips to the frozen north, retrieval of the frozen corpses of victims in Alaska. I won't spoil it.[18] According to Taubenberger's analysis, the H1N1 virus was *entirely* avian in origin rather than what's known as a *reassortant* virus, which results from mixing and matching of segments of already existing human flu viruses. This is compatible with the virus having undergone mutation in a mammalian intermediate host that made it more amenable to human infection.[19]

In 1918, discovery of the viral basis of flu in general was still 15 years away. And the kind of gene-sequencing technology required to demonstrate specific animal origin of an individual flu variant was, of course, much further away than that. We now know that all human and mammalian flu viruses ultimately have an avian origin. Specifically, they originate in ducks. Even more specifically, in the intestines of ducks.[20] As long as the viruses stay there, they remain largely harmless. It's only when they move around, and perhaps mingle, that they become deadly. Before, humans started domesticating ducks, around 4500 years ago, there almost certainly was no such thing as human influenza.[21]

Can't Knock the Hustle

There is a song by the band *Weezer*, a tribute to late 1970s funk music, that, I think, quite accurately captures the essence of the influenza virus. The song is called "Can't Knock the Hustle," and it begins thus:

[17] J. Taubenberger, A. Reid, R. Lourens, R. Wang, G. Jin, and T. Fanning, "Characterization of the 1918 Influenza Virus Polymerase Genes," *Nature* 437 (2005): 889.

[18] The CDC website relates the story here: https://www.cdc.gov/flu/pandemic-resources/reconstruction-1918-virus.html.

[19] As we shall see in the next section, mutation and reassortment are quite different. Taubenberger's denial of reassortant character has been contested. See, for example, M. Gibbs and A. Gibbs, "Was the 1918 Pandemic Caused by a Bird Flu?" *Nature* 440 (2006): E8, . https://doi.org/10.1038/nature04823. For my purposes, this does not matter, since, on Gibbs and Gibbs view, the reassortment in question is not that of a prior *human* flu virus.

[20] J. C. Cohen and W. G. Powderly, *Infectious Diseases*, 2nd ed. (London: Mosby, 2004). Citation due to Greger, *How to Survive a Pandemic*, 431.

[21] K. Shortridge, "Severe Acute Respiratory Syndrome and Influenza," *American Journal of Respiratory and Critical Care Medicine* 168 (2003): 1416–1420.

> My manager's slacking so I gotta' move quick
> I'm lacking in natural gifts
> I'm an ugly motherfucker but I work hella' harder
> And you can write a blog about it

Many blogs and websites have been devoted to the Spanish flu, of course, and I suppose the ugly motherfucker bit is self-explanatory. *Lacking in natural gifts*, however, is genuinely apposite. Flu is an RNA, rather than DNA, virus. All RNA viruses are lacking in natural gifts in the sense that they can't do very much on their own. They can't move under their own volition. They don't, for example, have those useful *flagellae* that some bacteria have to propel themselves in the direction of greater mischief. Instead, the virus must infect host cells and get those cells to do all the work for it. The work in question is that of reproducing the flu virus. Once it gets inside a cell of an infected organism, it transforms the cell into a virus-producing factory. First, it slices up the cell's DNA, rewiring the cell to produce more of the virus RNA. The cell, essentially, becomes a drone, reproducing the virus's RNA with a devotion that will eventually lead to its own death—by which time, the virus will have spread to further cells. "Hasta luego, adios!" as *Weezer* also mentioned in the chorus.

"My manager's slacking." Well, if you work in the publishing industry and your manager happens to be a proofreader or copyeditor, this is also not far from the truth, although, admittedly, in this case it's not so much that the manager is slacking as nonexistent. As an RNA virus, influenza comprises no more than seven or eight RNA segments wrapped in a protein sheath. Any RNA virus is a *master of metamorphosis.*[22] When cells built from DNA replicate, they do so under the guidance of a zealous copyeditor, who assiduously checks for errors in copies and eliminates them wherever possible. A natural proofreader is, in this sense, built into DNA cells. No such proofreader exists in structures—including many viruses—built from RNA. These are less *Encyclopedia Britannica* and more *Wikipedia*. When a virus infects an organism, it will go about churning out copy after copy of itself. But these copies are often inexact replicas. In this way, millions of viral offspring are produced, each one an imperfect replica of the one that came before it. Many of these copies will go to the wall, crippled through ill-advised mutation. But some will not. Some will thrive. And when they do, they may go on to

---

[22] J. De Jong, G. Rimmelzwaan, R. Fouchier, and A. Osterhaus, "Influenza Virus: A Master of Metamorphosis," *Journal of Infection* 40, no. 3 (2000): 218–228.

become the dominant strain of a virus. Until a better mutation comes along at least. Nothing in nature mutates faster than an RNA virus—although some of these mutate fast than others. What, for larger creatures, would require millennia of evolution can be achieved by an RNA virus in little more than an afternoon. In short, RNA viruses are astonishingly sloppy at reproducing themselves. And that is why they are so successful: through an irresistible combination of hard work and dumb luck. Can't knock the hustle, indeed.

It's this kind of hustle that explains why there is no one-time-fixes-all vaccination for flu and other RNA viruses like HIV. Their sloppy replication results in what is known as *antigenic drift*. In the case of flu, this antigenic drift will cause the shape of the virus's outer surface—specifically, the arrangement and disposition of the hemagglutinin and neuraminidase spikes—to change, making the descendant of a flu virus unrecognizable, or only partially recognizable, to antigens that were formed to dismast one of its ancestors.

However, mutation leading to antigenic drift is not the only problem posed for our health by the flu virus. It is not even the most serious problem. The most serious problem is the flu's proclivity for *hybridization*. The genes that make up a flu virus are distributed between seven or eight different strands of RNA. When a flu virus takes over a cell, it will make millions of copies of these seven or eight different RNA strands. However, if two flu viruses take over the same cell at the same time, then these viruses can mix and match segments of their RNA. Essentially, it's the closest viruses ever come to having sex. The result can be the creation of a new hybrid virus. For example, the 2009 swine flu, or swine-origin influenza A—also a version of the H1N1 strand—was a hybrid of this sort. Six of its genes approximate in sequence to a family of H1N2 "triple-reassortant" influenza found in pigs in North America around the turn of the twenty-first century. Two further genes derive from Eurasian "avian-like" pig viruses. One of these is the NA gene, which closely resembles human H1N1 viruses identified in Europe in 1991–1993. The other is the MP gene, and this most closely resembles H3N2 viruses found in Asia in 1999–2000.[23]

What do you need for two different flu viruses to enter a host cell at the same time? Well, most straightforwardly, this will happen when the host animal is infected with both viruses at the same time. For example, if a worker

---

[23] A. Gibbs, J. Armstrong, and J. Downie, "From Where Did the 2009 'Swine-Origin' Influenza A Virus (H1N1) Emerge?" *Journal of Virology* 6, no. 207 (2009). https://doi.org/10.118.

in a chicken facility struggles into work while suffering from a human influenza strain and, on that day, picks up a form of avian influenza from one of the chickens, the stage has been set for this type of encounter and potential gene-swapping. There are, however, other possibilities.

Old MacDonald Had a Biohazard Facility

Whenever two or more flu viruses meet, there is always the possibility of this kind of reorganization of RNA sequences and the creation of a new hybrid virus. Because of the volatile nature of flu viruses, their proclivity for rapidly churning out, often imperfect, copies of themselves, this kind of reassortment of genetic material need not take long at all. Viruses get together when hosts of those viruses get together—or, more typically, *we bring them together*. Today, raising of animals for food is the principal way in which the influenza virus gets to mingle and mix. I say "today," but it's always been this way, ever since we started raising different animals for food in fact. It's just that, today, things are worse than ever before.

As a result of their rapid mutation, flu viruses are adept at recycling themselves through different hosts. All flu viruses originally hail from ducks, but from ducks they can pass to other birds. The jump to humans and other mammals is more difficult. Flu viruses use sialic acid receptors to lock on to host cells. However, cells in the intestines of ducks have linkages of a certain specific sort—known as *alpha-2,3 linkages*. It used to be thought that human lung cells do not have sialic acid receptors with these alpha-2,3 linkages, instead having only *alpha-2,6 linkages*. If this were correct, it would be difficult for an intestinal duck virus that specialized in alpha-2,3 linkages to set up shop in the human lung, and the passing of a flu virus directly from a duck to a human would be unlikely at best. Now, however, it seems, the picture is more nuanced. While the sialic acid receptors in human lung cells are dominated by alpha-2,6 linkages, alpha-2,3 linkages do exist in human lung cells in far smaller numbers.[24] This means that directly inheriting a flu virus from a duck is certainly possible. However, it also means that it is far easier for a human to acquire a virus that specializes in alpha-2,6 linkages.

---

[24] M. Matrosovich, T. Matrosovich, T. Gray, N. Roberts, and H. Klenk, "Human and Avian Influenza Viruses Target Different Cell Types in Cultures of Human Airway Epithelium," *Proceedings of the National Academy of Sciences of the United States of America* 101, no. 13 (2004): 4620–4624.

Enter the humble pig. In the lungs of pigs, both alpha-2,3 and alpha-2,6 linkages are both relatively widespread. This makes them vulnerable both to avian flu and its human counterpart, making them ideal vessels in the hybridization of both forms of influenza virus. Moreover, it means they are an ideal intermediate host from birds to humans.[25] And sometimes, also, back again—this mixing of influenza strains can be bidirectional. A virus that begins with a bird can makes its way into a pig and then be cycled back into the bird in a form that, due to the time spent adjusting to pig biology, is more readily able to infect mammals directly. From there it may now be able to pass to a suitable mammal—for example, one that spends a lot of time in the vicinity of the bird in question. In such cases—and this may be what happened with the Spanish flu—pigs play a role in the development of the virus, but they need not be directly involved in the transmission of the virus when that virus takes off in human beings.

Given these factors, then, a good rule of food production would seem to be this: *Don't let the pigs anywhere near the chickens!* Or the turkeys, ducks, geese—don't let the pigs near any kind of fowl, really. As the old saying goes: *Pigs and fowl should not be cheek by jowl.* Okay, there probably is no such old saying. But there should be. Here, for example, is a just about perfect example of what not to do. You gather together tens of thousands of chickens and pigs. You place the chickens in cages above the pigs. And you allow them to defecate on the pigs below. These chicken droppings actually form part of the diet of the pigs—you have designed it this way to save money. Genius! Except . . . it is well known that influenza virus can survive in chicken poop for weeks. If the chickens have an avian virus, then those alpha-2,3 sialic acid receptors of the pigs are going to do their work. But no one would raise pigs and chickens in this way, you think? On the contrary, it forms two-thirds of the fish-farming—or pig-hen-fish aquaculture—method widely employed in Asia. In this tripartite method, chickens and pigs are arranged in the way described here. But below the pigs are fish. The pig excrement is either directly eaten by the fish or is used as fertilizer for aquatic plant-based fish food (it depends on the species of fish).[26] The pond water can then be pumped back up to be used as drinking water for the chickens and pigs.[27]

---

[25] T. Ito, J. Couceiro, S. Kelm, et al., "Molecular Basis for the Generation in Pigs of Influenza A Viruses with Pandemic Potential," *Journal of Virology* 72, no. 9 (1998): 7367–7673. See also Greger, *How to Survive a Pandemic*, 66.

[26] Greger, *How to Survive a Pandemic*, 160–162.

[27] S. Matsui, "Protecting Human and Ecological Health under Viral Threats in Asia," *Water Science and Technology* 51, no. 8 (2005): 91–97.

This kind of integrated aquaculture is, fiscally, very efficient, offering significant reductions on feed and waste disposal costs. As such, it has attracted increasing support from international aid agencies.[28]

The ultimate source of the flu virus in chickens will be wild ducks. Integrating chicken, pig, and fish farming allows the virus the wonderful opportunity to cycle through birds into mammals and back into ducks. This is an opportunity we have engineered through our farming practices, and it is not difficult to see how it might go. Virus-laden migratory ducks land on a farmed fish pond. The virus, lying harmlessly in the ducks' intestines, is excreted into the water. The chickens drink the water and acquire this avian virus. The pigs also drink this water. They eat the chickens' feces and are also fed the remains of dead chickens from the facility. The virus continually reassorts itself, rearranging segments of RNA. Eventually, through dumb luck and hard work—can't knock the hustle—a new RNA virus strain is produced, one capable of setting up home in a pig. The pig distributes this new, hybrid virus in the ponds—integrated plumbing and waste disposal not a strong point of these sorts of farms. Here, it may contain enough of its original avian characteristics to reinfect the ducks. The migratory paths of the ducks, then, distribute the virus thousands of miles, infecting other ducks, chickens, and pigs.

As long as the virus stayed in the ducks—where viruses like it have resided for millennia and lived entirely unremarkable lives—there would be no pandemic threat. Bringing chickens into the picture changes matters. A virus that is harmless to ducks is often anything but harmless when it makes the jump to chickens. Sometimes this will result merely in less than perfect health, reducing weight gain—a factor that the chicken farmer will have to take seriously. But, sometimes, as I mentioned in the previous chapter, the jump to chickens seems to positively weaponize the virus. This happened with the outbreak of the H5N2 virus that struck chicken farms in Pennsylvania in 1983. In total, 17 million chickens either died from the virus or had to be culled to prevent its further spread. Necropsies showed a variety of congestive, hemorrhagic, and necrobiotic causes of death.

Most famously—so far—there was the Hong Kong outbreak of the H5N1 virus in 1997, which exhibited similar weaponization of a formerly harmless duck virus. In the spring of 1997, thousands of chickens began dying—suddenly and nastily—from a H5N1 virus. Sometimes chickens would die

---

[28] Greger, *How to Survive a Pandemic*, 160.

of asphyxiation, choking on large clots of blood lodged in their windpipes. Necropsies sometimes revealed that the chickens' internal organs had been reduced to a bloody pulp.[29] Then people started dying. The first was a 3-year-old boy, who died of multiple internal organ failure and coagulation of the blood. The second was a 13-year-old girl who died of internal hemorrhaging. The initial symptoms were typical of a flu: fever, headache, muscles aches, malaise, congestion. The virus then went on to attack internal organs, via the kind of cytokine storm that has become all too familiar from COVID-19. Most patients died of multiple organ failure and/or drowning on blood-filled lungs.[30] In the 1997 outbreak, the virus infected 18 people, 6 of whom died. The common factor uniting all individuals was direct or indirect contact with commercial poultry. The outbreak was eventually halted in December 1997 with the culling of all of Hong Kong's 1.5 million chickens.[31]

The mercifully limited human death toll in this outbreak was due to the fact that H5N1 had not yet mutated into a form that allowed for easy human-to-human transmission. However, limited human-to-human transmission did occur through close physical contact. Despite this setback in Hong Kong, the H5N1 virus itself thrived and spread. By the end of 2003, it had spread beyond China to encompass much of Asia. From there, almost certainly facilitated by a hemispheric trade in live birds, and perhaps also by wild bird migrations,[32] it spread to the Middle East, Europe, and Africa.[33] Within a few months of 2004, more than 100 million birds were dead, either from the virus or through culling. The year 2005 saw a resurgence in human deaths. Over 100 laboratory confirmed cases were recorded, with more than 50 deaths. Now H5N1 is endemic in wild bird populations in Asia, Africa, Europe, and the Middle East, and semi-regularly pops up in domestic bird populations. There is no putting that particular genie back in the bottle.

In this context, the practice of raising pigs anywhere near the vicinity of chickens seems positively suicidal. Pigs, as we have seen, are ideal vehicles for an aspiring flu virus, eager to expand its sphere of influence, to practice on mammals: to gradually become acclimated to mammalian biology, to

---

[29] Greger, *How to Survive a Pandemic*, 59.

[30] P. Chan, "Outbreak of Avian Influenza A(H5N1) Virus Infection in Hong Kong in 1997," *Clinical Infectious Diseases* 34, no. 2 (2002): S58–S64, https://doi.org/10.1086/338820.

[31] Chan, "Outbreak of Avian Influenza A(H5N1) Virus Infection in Hong Kong in 1997."

[32] Y. Si, A. Skidmore, T. Wang, W. de Boer, P. Debba, A. Toxopeus, L. Li, and H. Prins, "Spatio-temporal Dynamics of Global H5N1 Outbreaks Match Bird Migration Patterns," *Geospatial Health* 4, no. 1 (2009): 65–78.

[33] Editorial, "Avian Influenza Goes Global, but Don't Blame the Birds," *Lancet Infectious Diseases* 6, no. 185 (2006), list.web.net/archives/sludgewatch-l/2006-April/001692.html.

learn to how to cope with a mammalian bodily host, and perhaps eventually to spread from one such host to another. It is true that aquaculture is not a particularly common way to farm pigs and chickens, perhaps involving only 20% of pigs farmed in Asia.[34] Nevertheless, that is still tens of millions of pigs. Moreover, and this is crucial, there are multiple opportunities for virus recombination and resulting hybridization in farming practices that are much more widespread, including in the United States. Remember, the golden rule is, *pigs and fowl should not be cheek by jowl*. Unfortunately, this rule is routinely broken. It is increasingly common to find pig and chicken facilities standing side by side with each other.[35] Intensive facilities such as these, although labeled "confined," are hardly hermetically sealed. Air has to be cycled in and out, and the exhaust fans can theoretically transfer virus particles from one facility to the next.[36] Chicken droppings might be brought in on the foot of a wandering mouse or rat; or pig feces taken in the other direction by the same means.

More importantly, pig and chicken facilities often share common workers—that is one of the financial attractions of having the facilities side by side. That and being able to easily feed dead chicken to the pigs. Humans played an essential role in the swine flu epidemic of 2009, which killed half a million people. This was a, hitherto unprecedented, triple reassortant virus, where the classic swine flu virus acquired three gene segments from a circulating human flu virus, and two gene segments from an avian flu virus, to form this novel version of swine flu.[37] This novel variant was first detected in 1998, at an industrial pig operation in Newton Grove, North Carolina.[38] Within a year, it had spread to pig facilities across the United States. This rapid spread was the result of the ubiquitous long-distance travel to which many intensively raised pigs are subject—another respect in which confined operations are far from hermetically sealed. A pig can, for essentially commercial

---

[34] P. Edwards, "Fish-Fowl-Pig Farming Uncommon," *World Aquaculture* 22, no. 3 (1991): 2–3.

[35] Webster and Hulse decry this "recently evolving intensive farming practice in the USA." R. Webster and D. Hulse, "Microbial Adaptation and Change: Avian Influenza," *Revue scientifique et technique* 23, no. 2 (2004): 453–465, https://doi.org/10.20506/rst.23.2.1493.

[36] J. Graham, J. Leibler, L. Price, J. Otte, D. Pfeiffer, T. Tiensin, and E. Silbergeld, "The Animal-Human Interface and Infectious Disease in Industrial Food Animal Production: Rethinking Biosecurity and Biocontainment," *Public Health Report* 123, no. 3 (2008): 282–299, https://doi.org/10.1177/003335490812300309.

[37] Greger, *How to Survive a Pandemic*, 164.

[38] The Smithfield-owned Granjas Carroll pig facility in La Gloria, Mexico, is often identified as ground zero, since it was the location of the first known human casualties. But the swine flu virus was in existence and spreading some time before this human outbreak.

reasons, be born in one state, fattened in another, and slaughtered in yet another.[39]

Days of Future Past

There are several parallels between the Spanish flu (H1N1) and the H5N1 virus. Both are unusually lethal, although H5N1, so far, is an order of magnitude worse, having killed around 50% of humans who have been unfortunate enough to get it.[40] Both also often induce death through the kind of *cytokine* storm made famous by COVID-19, where the immune system of victims becomes hyperactive, killing its host. In the 1997 H5N1 outbreak, autopsies of victims revealed very high cytokine levels. Interestingly, viral cultures taken during autopsy came up negative, showing that these cytokine storms had, in one sense, triumphed: cleansing the body of the H5N1 virus. A Pyrrhic victory, of course: in destroying the virus, the immune system had also destroyed the host. Recorded observations of the manner of death of victims of the Spanish flu also suggest a prominent role for cytokine storms. This is supported by the fact that both the Spanish flu and H5N1 tend to take those in the prime of life rather than, as in the case of seasonal flu, the very old and the very young.

The lethality of both the Spanish flu and the H5N1 virus might also be the result of a trick that each virus is capable of pulling off that hides its replication from the host's immune system. The body's immune response to a flu virus involves a group of signal proteins known as *interferon* and an enzyme known as PKR. When an RNA replicates inside a host cell, there will be momentarily, two RNA strands contained inside the cell. In such circumstances, interferon proteins signal to neighboring cells to produce the PKR enzyme, and the function of the latter is to stop protein synthesis in the cell—which both stops replication of the virus and kills the cell. However, H5N1 has a trick up its sleeve: NS1—nonstructural protein—which binds to the double RNA strand and, effectively, hides it from PKR. NS1 covers up the tell-tale sign of the virus. All influenza viruses have NS1, but H5N1 seems to have a mutant form of this protein that is much more effective in hiding the twin

---

[39] Greger, *How to Survive a Pandemic*, 16.

[40] World Health Organization, "Cumulative Number of Confirmed Human Cases for Avian Influenza A(H5N1) Reported to WHO, 2003–2020," https://www.who.int/influenza/human_animal_interface/2020_01_20_tableH5N1.pdf.

strands of RNA.[41] Essentially, the flu virus used NS1 as a way of buying time to produce enough of itself. When eventually discovered, the host's immune system may "panic" and call in all the troops, resulting in a cytokine storm.

At present, H5N1 is a stage 3 pathogen—in the sense explained in the previous chapter. That is, it can be transmitted to humans from birds, and some human-to-human transmission has occurred. It does not, however, spread easily between humans. This is fortunate. With a mortality rate in humans of around 50%, if the virus ever does acquire the ability to move easily between humans, while retaining this level of virulence, the death toll would be terrifying.

With this potential death toll in mind, the trajectory of H5N1 since the Hong Kong outbreak is a little concerning. In addition to becoming endemic in wild and domestic bird populations in many parts of the world, perhaps most worrying of all are the inroads the H5N1 virus has been making into mammal populations in the past two decades. By 2003, H5N1 had undergone almost a dozen reassortments with other influenza viruses. The result was a hybrid H5N1 virus known as type Z. By 2005, this had undergone further hybridization to become type Z+. H5N1 Z+ is both incredibly lethal and incredibly infectious. The latter feature allowed H5N1 to spread uncontrollably among birds across Asia and the Middle East. The former feature—lethality—is not confined to birds. In a University of Wisconsin study, H5N1 Z+ killed 100% of a test population of mice, effectively dissolving their lungs, causing the researchers to describe it as the "most pathogenic virus we know of."[42]

In the decade and a half it has existed, the Z+ variant of the H5N1 virus has been steadily expanding the number of mammals it can infect. There have been reports of infected pigs in China.[43] As we have seen, this is particularly worrying, given what we know about the ability of pigs to catch both human and avian influenzas, and the resulting possibilities for hybridization of the viruses that meet up in this way. Some pet cats have died.[44] This is the first time on record of cats catching influenza—prior to this they were thought of as immune to these viruses. Tigers and leopards in zoos that were

[41] Greger, *How to Survive a Pandemic*, 46.

[42] M. Drexler, *Secret Agents: The Menace of Emerging Infections* (Washington, DC: Joseph Henry Press, 2002), 180.

[43] D. Cyranoski, "Bird Flu Data Languishing in Chinese Journals," *Nature* 430 (2004): 995.

[44] World Health Organization, "Avian Influenza A(H5N1)—Update 28: Reports of Infection in Domestic Cats (Thailand), Situation (Human) in Thailand, Situation (Poultry) in Japan and China," February 2004, who.int/csr/don/2004_02_20/en/.

fed infected chickens have also died of H5N1.[45] Worryingly, there is also evidence of the virus being spread from tiger to tiger.[46] There is also a case of H5N1 infection of a dog.[47]

At present, however, while the H5N1 virus is transmissible to mammals, it does not spread easily between mammals. To become readily transmissible between mammals, either some kind of *mutation* or some kind of *reassortment* will be required. Herfst and colleagues set about estimating the number of mutations required to allow for aerosol transmission between humans and arrived at the surprisingly, and very worryingly, low figure of five.[48] With regard to reassortment, Imai and colleagues have provided a model for how certain, relatively minimal quantities of reassortment could result in a version of H5N1 that is transmissible between mammals.[49] Some have been encouraged by the failure of H5N1 to, as yet, transform into a virus capable of mammal-to-mammal transmission. However, one might look at H5N1 and see a virus gradually increasing its familiarity and facility with mammals. Every step it takes in this process makes it more likely to evolve in a stage 4 pathogen—shaping it into a form readily transmissible between humans. If that happens, we are back in 1918. Or worse.

High-Risk Behavior

When the AIDS epidemic was in full swing in the 1980s, much was made of the need to eschew *high-risk behaviors*. No sex without condoms. No reusing hypodermic needles for one's heroin infusion. High-risk behaviors were at the core of HIV transmission, and once this was understood, people did get much better at moderating their high-risk behavior. Today there is one form of high-risk we can't persuade ourselves to eschew—perhaps because we don't, in general, even recognize that it is a high-risk behavior. Raising animals for food, especially in the way we do it—intensively—is perhaps the

---

[45] J. Keawcharoen, K. Oraveerakul, T. Kuiken, et al., "Avian Influenza H5N1 in Tigers and Leopards," *Emerging Infectious Diseases* 10 (2004): 2189–2191.

[46] T. Kuiken, G. Rimmelzwaan, D. van Riel, et al., "Avian H5N1 Influenza in Cats," *Science* 306 (2004): 241.

[47] CDC, "Influenza," https://www.cdc.gov/flu/avianflu/h5n1-animals.htm.

[48] S. Herfst et al., "Airborne Transmission of Influenza A/H5N1 Virus between Ferrets," *Science* 336 (2012): 1534–1541.

[49] M. Imai, T. Watanabe, M. Hatta, et al., "Experimental Adaptation of an Influenza H5 HA Confers Respiratory Droplet Transmission to a Reassortant H5 HA/H1N1 Virus in Ferrets," *Nature* 486 (2012): 420–428. https://doi.org/10.1038/nature10831.

most serious form of high-risk behavior in which humans, today, willingly engage.

You might think of the matter in this way. The Centers for Disease Control and Prevention (CDC) classifies H5N1 as a ("at a minimum") Biosafety Level 3 pathogen.[50] This means that the virus is only to be handled in high-containment buildings, hermetically sealed with air locks. Access to individual corridors must be controlled and employ double-door entries. Access is limited to highly trained personnel. Showering is required upon every entry and exit. Air flow is unidirectional with a filtered exhaust. Floors, walls, and ceilings are waterproofed. All wall penetrations, such as electrical outlets, phone, and Internet lines, are caulked and sealed. Surfaces are disinfected on a daily basis, and solid wastes are incinerated.[51]

Let's contrast this with a confined animal feeding operation (CAFO) or, more colloquially, a factory farm, which will have none of these measures. An operation that will be continually bringing new animals in and shipping old ones out, perhaps traveling hundred or even thousands of miles on the road to fattening and slaughter. But we know viruses such as H5N1 are endemic. They pop up, with depressing regularity, in such operations, leading to the slaughter of millions of birds. Imagine if some scientists got together, at one of their houses, and decided to play around with some Biosafety Level 3 pathogens in the garage. When play time is over, the rest of them go back to their own houses, and hug their families, and sit down to dinner with them.

In effect, this is exactly what is going on in CAFOs. If circumstances appropriately align, down on the old factory farm, we will find ourselves trying to deal with a Biosafety Level 3 pathogen in a Biosafety Level 0 facility. Risky. And every day it gets just a little riskier. As the meat industry intensifies—as more and more of the chickens we eat are raised in CAFOs—so do the outbreaks of *highly pathogenic avian flu* (HPAI) correspondingly increase. There was one outbreak of HPAI in the 1950s, two in the 1960s, three in the 1970s, three in the 1980s, and nine in the 1990s. Then H5N1 came along, and since then there have been 6500 outbreaks in more than 60 countries.[52]

---

[50] CDC, "Work with HPAI H5N1 Virus Should Be Conducted, at a Minimum, at Biosafety Level 3 (BSL-3), with Specific Enhancements to Protect Workers, the Public, Animal Health, and Animal Products," https://www.cdc.gov/mmwr/preview/mmwrhtml/rr6206a1.htm#:~:text=Work%20 with%20HPAI%20H5N1%20virus,animal%20health%2C%20and%20animal%20products.

[51] Description due to Greger, *How to Survive a Pandemic*, 371–372.

[52] M. Van Kerkhove, S. Vong, J. Guitian, D. Holl, P. Mangtani, S. San, and A. Ghani, "Poultry Movement Networks in Cambodia: Implications for Surveillance and Control of Highly Pathogenic Avian Influenza (HPAI/H5N1)," *Vaccine* 27 (2009): 6345–6352, https://doi.org/10.1016/ j.vaccine.2009.05.004.

According to Food and Agriculture Organization (FAO) statistics, the three most commonly slaughtered animals in the world are chickens, ducks, and pigs. Pig and fowl, cheek by jowl: from the perspective of future HPAI pandemics, this is as dangerous a combination as one can get.

Of course, maybe it won't be H5N1 that gets us. Maybe it won't be another influenza virus at all. Another coronavirus, perhaps? With higher mortality and infectivity than COVID-19? Or maybe it will be a filovirus? Another retrovirus, killing tens of millions, perhaps? Or maybe a paramyxovirus, like Nipah? I have to admit, Nipah worries me a bit. Indeed, the trajectory of Nipah since the first outbreak in 1998 is, in its own way, at least as worrying as that of H5N1.[53] After Malaysia, Nipah moved on to Bangladesh and, there, outbreaks have been recorded in most years since 2001. There are several things that are worrying about this. There are the cases of human-to-human transmission, which are becoming quite frequent. Indeed, in Bangladesh, human-to-human transmission now accounts for one-third of cases. In one instance, the virus was even acquired from a corpse, which suggests a quite high level of infectivity. Then, there is the case fatality rate of 70%, which speaks for itself. Just as worrying is the evident level of mutation the virus is capable of undergoing. The Malaysian strain of Nipah presents quite differently from the Bangladesh strain. In Malaysia, only 14% presented with a cough. In Bangladesh, 62% of patients presented with cough. In Malaysia, only 6% of patients had abnormal chest radiographs. In Malaysia, 69% of patients had abnormal chest radiographs, demonstrating substantial pulmonary damage. These differences in the way the strains present may be responsible for the difference in human-to-human transmission rates in Malaysia and Bangladesh. At present, even in Bangladesh, Nipah is still a stage 3 pathogen, with limited human-to-human transmission, and an estimated R_0 of 0.48.[54] However, it is clear that Nipah is a highly mutable virus. It is conceivable that one of the numerous strains of Nipah circulating in Bangladesh has already achieved a R_0 of greater than 1.[55] Worries about the infectivity of

---

[53] The following discussion is indebted to S. Luby, "The Pandemic Potential of Nipah Virus," *Antiviral Research* 100, no. 1 (2013): 38–43.

[54] Luby, "The Pandemic Potential of Nipah Virus."

[55] Thus claims Luby, "The Pandemic Potential of Nipah Virus," Section 7: "The combined evidence of genetic heterogeneity of virus strains in Bangladesh, the different transmission pattern seen in outbreaks with different strains and the ferret studies suggests that different strains of Nipah virus possess different capacities for person-to-person transmission. It is conceivable that one of the numerous strains circulating in bats throughout Asia may have an R_0 value >1.0 for humans or acquire mutations during human infection that lead to more efficient and sustained human-to-human transmission."

Nipah are not simply idle, or paranoid, speculation. It has happened before with another, far more familiar, paramyxovirus: measles. Measles apparently originated in the eleventh and twelfth centuries, a mutant form of the rinderpest virus, hosted by ruminants.[56] This variant strain of paramyxovirus evolved a remarkably high R_0 (c. 18) and went on to kill tens of millions of people in subsequent centuries.

So, maybe Nipah? Maybe H5N1? Maybe it will be this virus, maybe it will be that one. Maybe it will be one we don't know yet. But we can virtually guarantee that some or other zoonotic virus is on its way, and how bad it is going to be we will just have to wait and see. We can guarantee that some or other zoonotic virus will *always* be on the way—unless it has actually already arrived—as long as we continue to engage in the high-risk behavior of eating animals. If we don't have to do it—and we don't—eating animals is *insane*. One day—not today, perhaps not tomorrow, but perhaps sooner than we think—this collective insanity may, as it did 100 years ago, kill tens of millions of us. Or more.

[56] Y. Furuse, A. Suzuki, and H. Oshitani, "Origin of Measles Virus: Divergence from Rinderpest Virus between the 11th and 12th Centuries," *Journal of Virology* 7 (2010): 52.

12

The Wildwood

Stopping and Starting

Our tenure on this planet has coincided with the most astonishing biomass reallocation program imaginable. Mammalian biomass—96% us and the mammals we farm. Avian biomass—70% domestic fowl. That does seem rather, shall we say, *slanted*. This colossal biomass reallocation program has been attained only through the relentless confiscation of vast expanses of land. We continue to take more and more of the ancestral lands of animals whose biomass we value less. This hunger for land is driven by our hunger for meat. And the result? *Climate change*, *mass extinction*, and *pestilence*.

To combat these threats, our task is nothing less than reversing the arc of human history, an arc defined and delimited by this reallocation of biomass. Specifically, we need to *stop* doing something and *start* doing something else. We need to *stop eating animals*. This is a habit, not an addiction. And certainly not a necessity. At one time, it may be that this habit served us well. Perhaps, by giving us access to high-quality, high-energy sources of food, it allowed us to do things—such as grow bigger brains—that were essential to our development as anatomically modern human beings.[1] Perhaps. But times have changed. Food has changed, and there are many more sources of energy and nutrition available to us now. We no longer need to eat animals and can, if we so desire, move on. If we are serious about the environmental threats we now face, we *must* move on.

No longer eating animals, breaking this most ancient, most tenacious, and ultimately most pernicious habit will, on its own, accomplish two things. It will, if the Intergovernmental Panel on Climate Change (IPCC) is to be believed, subtract 14.5% of the CO_2-equivalent greenhouse gas (GHG)

---

[1] Richard Wrangham has pursued this line of argument in *Catching Fire: How Cooking Made Us Human* (New York: Basic Books, 2009). As a historical observation, I have no issue with this argument. It may also explain just why many of us—me included I seem to remember—find meat so very *tasty*. We have evolved to do so. But, as a prescription for what course of action we should adopt now, this observation, of course, has no weight. The inference from "We needed to do this once" to "Therefore, we need to do it now" is, of course, an invalid one.

World on Fire. Mark Rowlands, Oxford University Press. © Oxford University Press 2021.
DOI: 10.1093/oso/9780197541890.003.0012

emissions we send into the air every year. Eating animals is, fundamentally, an energy exchange that has gone badly awry. By eschewing this habit, simplifying and making more efficient the energy supply train that keeps us and our societies in existence, we take some of the pressure off beleaguered renewable energy sources. There is, I suspect, ultimately no alternative to renewables, but the energy returned on energy invested values (EROIs) of renewables are too close to the cusp of viability—tottering near the edge, and staring out into the abyss, of the net energy cliff. By excising, from the energy-in side of the equation of our lives, the strangely inverted EROIs of meat and other animal products, we give our renewable energy sources time to bed in, and their EROIs—hopefully, but it's a reasonable hope—time to begin ticking upward.

The other thing we immediately do, when we abandon our old habit, is insulate ourselves from the reservoirs of zoonotic pathogens that are found in the animals we eat, both wild and domestic. Most of the diseases that have infected and afflicted us have been zoonotic—zoonoses or of zoonotic origin. Almost all newly emerging infectious diseases are zoonotic. In eating animals, we engineer opportunities for these diseases to thrive. We do so by bringing animals and their diseases together. We bring bats together with civets and pangolins. We bring ducks together with chickens, and chickens together with pigs. And, all the while, pathogens mix and mingle. Sometimes the results are unremarkable. Sometimes they are calamitous. This mixing and mingling, mutation and reassortment, of pathogens is driven by our hunger for meat. A bat may impart Nipah virus to a pig. But only a human can make that pig into an intermediate host. A few small exceptions aside, if we no longer eat animals, the intermediate hosts of zoonotic viruses immediately disappear. We thereby shield ourselves from vast reservoirs of potential pathogens—harmless in their original hosts, often anything but that in us.

The third thing we do is by breaking our old habit is immediately arrest some of the more important drivers of species extinction currently in play on the earth and in the oceans. According to the Intergovernmental Science-Policy Platform on Biodiversity and Ecosystem Services (IPBES), direct exploitation of animals—colloquially known as hunting and eating them—is the second most important driver of terrestrial animal extinction. Turning to marine extinctions, no longer eating animals will eliminate the number-one cause of these: direct exploitation of marine animals through hunting. If we stop eating animals, we straightforwardly take these drivers of extinction out of commission.

However, I have not even got to the most important benefit of breaking our accursed and ancient carnivorism. It is not what this immediately does but what it makes possible that is most important. If we can break our predilection for meat, we thereby break our ravening hunger for land. I mentioned that to properly combat the threats we face, we need to stop doing something and start doing something else. We stop eating animals. And we thereby start—we have made this possibility available—*afforesting wherever and whenever we can*. This is where the magic happens.

The Magic Bullet?

As we saw in Chapter 7, Jean-Francois Bastin and colleagues estimate that under current climatic conditions, there is an additional 2.2 billion acres of land suitable for afforestation, an area of land that, they estimate, could store around 205 gigatons of carbon.[2] Commenting on this paper in *The Atlantic*, Robinson Meyer writes: "A recent high-profile study, for instance, cheerfully suggested soaking up most of the planet's carbon pollution by planting 1.2 trillion trees across 2.2 billion acres worldwide. It was impressive research, but it was immediately sold to the public as The Solution to Climate Change. And there's a problem with that. Those 2.2 billion acres—an area roughly the size of the continental United States—*are already in use*. They comprise, in large part, the planet's most productive farmland!" Bastin and colleagues did, in fact, attempt to exclude land currently used for agriculture from their calculations. However, some critics have contended that their exclusion wasn't sufficiently comprehensive. In a letter to the journal *Science*, Eike Luedeling and colleagues contend that Bastin's estimates of land available for afforestation are too high. These estimates, they claim, include land currently used for grazing as well as some urban areas, and have also not adequately considered factors such as soil degradation, aridity, and other water constraints.[3] Bastin has responded to these criticisms.[4]

---

[2] J-F. Bastin, Y. Finegold, C. Garcia, D. Mollicone, M. Rezende, D. Routh, C. Zohner, and T. Crowther, "The Global Tree Restoration Potential," *Science* 365, no. 6448 (2019): 76–79.

[3] E. Luedeling, J. Börner, W. Amelung, K. Schiffers, K. Shepherd, and T. Rosenstock, "Forest Restoration: Overlooked Constraints," *Science* 366, no. 6463 (2019): 315.

[4] J-F. Bastin, Y. Finegold, C. Garcia, N. Gellie, A. Lowe, D. Mollicone, M. Sacande, B. Sparrow, C. Zohner, and T. Crowther, "Response to Comments on 'The Global Tree Restoration Potential,'" *Science* 366, no. 6463 (2019): eaay8108.

In this context of kind of technical disagreement about how much land, precisely, is available for afforestation, imagine if vast new swathes of what was formerly farmland, running into the billions of acres, were suddenly—*somehow*—made available for the afforestation project.

Much of this land—currently used for grazing, or growing animal feed crops, or the never-going-to-amount-to-anything biofuels—would, if your business is the growing and nurturing of new forests, be prime real estate. There is farmland we need and there is farmland we don't. A not inconsiderable portion of farmland—that used for grazing and much of that used for animal feed crops—is required as such only because of our dietary predilections. The land would cease to be needed for this purpose if these predilections were to appropriately change. The disastrous energy exchange that consists in the consumption of animals and their products commits us to using far more land than we need. If we abandon this consumption then—even allowing for the additional land we would need to allocate to growing crops for our consumption—we would make available huge amounts of fertile new land for afforestation.

Through the increased carbon sequestration that it affords, afforestation will allow us to significantly reduce our carbon footprint. By precisely how much is a technical debate, stymied, in part, by a lack of consensus on the carbon-sequestering potential of different types of forest. Nevertheless, the potential of afforestation for climate change mitigation is, of course, widely recognized, and has been for some time. It is the basis of a number of international initiatives, including the Bonn Challenge[5] and the New York Declaration on Forests.[6] Moreover, the latest report by the IPCC has recognized that at least 1 billion hectares of new forest (in addition to preservation of old forest) will be required in order to keep global temperature increases within 1.5°C—the aspirational goal of the Paris Agreement.[7]

It would, of course, be wrong to think of afforestation as *The Solution* to climate change. But if it is not *The Solution*, then it is, at the very least, a crucial part of *A Solution*. Fundamentally, what afforestation will do is *buy us time*: enough time for us to work on the EROIs of renewable energy technologies, and finally get GHG emissions under control. But afforestation will

---

[5] United Nations Environment Program (UNEP), The Bonn Challenge (2011) https://www.bonnchallenge.org/

[6] UN Climate Summit, New York Declaration on Forests (2014) https://forestdeclaration.org/

[7] Intergovernmental Panel on Climate Change (IPCC), *An IPCC Special Report on the Impacts of Global Warming of 1.5°C Above Pre-Industrial Levels and Related Global Greenhouse Gas Emission Pathways* (2019) https://www.ipcc.ch/sr15/

do this for us only in conjunction with the breaking of our ancient habit. Without this dishabituation, there simply won't be enough land to make it work. But, with it, the possibilities are exciting. Afforestation may not be a magic bullet in the attempt to slay climate change, but it is something that's really not too far off—a Magic-y bullet, perhaps.

Even more Magic-y, afforestation on this scale also helps solve the problem of mass extinction. It is our ravening hunger for land that, according to IPBES, is the most important driver of terrestrial species extinction. We rid ourselves of our hunger for meat, we thereby rid ourselves of our hunger for land, thereby taking out of play the most important driver of terrestrial species extinction. We also, as I mentioned earlier, take out the second most important driver of terrestrial extinctions and the most important driver of marine extinctions—direct exploitation of animals through hunting. Moreover, to the extent afforestation mitigates climate change, it also mitigates the third most important driver of both terrestrial and marine extinctions.

Third, afforestation on the scale envisaged helps further insulate us from animal reservoirs of zoonotic diseases. One of the most important reasons we come into contact with zoonotic pathogens is destruction of the environment that is home to the original hosts. A forest burns in Java. The forest's fruit bats flee to Malaysia, and soon we have an outbreak of Nipah virus to contend with. It is generally accepted that one of the most important steps we can take in the attempt to combat newly emerging infectious diseases is keeping these animal reservoirs safe and undisturbed in their ancestral lands.[8] The biggest change occurring in those ancestral lands today is deforestation. And the single biggest cause of this is deforestation performed in the service of the globally burgeoning meat industry. If we eschew our habit of eating meat, then we also eschew our hunger for this land. And this will allow us to maintain this land in a stable and undisturbed-as-possible state. In the fight against newly emerging infectious diseases with pandemic potential, afforestation is a crucial weapon in the armory.

It is our hunger for land that has been our undoing. In human portrayals of animal hunger, the wolf is often a central figure: the ravening wolf at the

---

[8] Since the appearance of COVID-19, there has been a slew of statements to this effect. See, for example, this essay, written by some collaborators on the next IPBES report, see J. Settle, S. Diaz, E. Brondizio, and P. Daszak, "COVID-19 Stimulus Measures Must Save Lives, Protect Livelihoods, and Safeguard Nature to Reduce the Risk of Future Pandemics," *IPBES Expert Guest Article*, https://ipbes.net/covid19stimulus. See also M. Everard, P. Johnston, D. Santillo, and C. Staddon, "The Role of Ecosystems in Mitigation and Management of Covid-19 and Other Zoonoses," *Environmental Science and Policy* 111 (2020): 7–17.

door. Little pig, little pig, let me come in. Let it go, Mr. Wolf. The pig has Nipah virus, and we pretty much engineered that, I'm afraid. Let it go. You could never compete with us, anyway. As far as hunger is concerned, it all comes back to us: our terrible hunger for land, born of our ravening hunger for meat.

The Big Picture

It is important that we always consider these three environmental problems together, and understand the connection between them, because if we did not, we would not be able to see the *big picture*. And if we cannot see the big picture, we would likely end up advocating ineffective, or positively harmful, solutions. For example, as we saw in Chapter 6, for those who are focused exclusively on climate change, the lesson we should take away from our examination of this problem is simple: *Eat mor' chikin*. We should eat chicken, instead of beef. Maybe pigs, too, but definitely chicken. This conclusion derives from an unduly narrow focus. First, even if the feed-conversion ratio (i.e., energy efficiency) of chickens is not as bad as it is for beef, it is still, in the great scheme of things, remarkably bad. Land must be used to grow chicken feed as well as cattle feed. Land that could be better used in other ways. Second, and most importantly, in the context of newly emerging infectious diseases, chickens, and also pigs, are the arch intermediate hosts: the animals that link us most closely to a viral reservoir of untold potential malignancy and consequences. Chickens may not be as bad as beef, energetically speaking. But, pathogenically speaking, it is an entirely different story.

Moreover, if we just focus on climate change, our conception of afforestation is likely to be—what's that word I used earlier?—*slanted*. A little parochial, let us say. If sequestration of carbon dioxide is our only concern, or our overriding concern, then we are likely to try to grow huge monoculture forests, with trees tightly packed, as many per acre, as they will grow. Hansel and Gretel would have had their asses handed to them by these forests, believe you me. The witch, too, probably. Less efficient forests, with trees more sparsely spread? We might even be tempted to destroy them and replace them with our chockablock monocultures. We shouldn't do this—obviously!—because there is not only a problem of climate change, but two others besides. And our tightly packed monoculture forests—where they do not belong—would do little to address these. Afforestation must be pursued with a solution to all three problems in mind. If it is not, the general outlines of the story are predictable. Bat meets virus. Human replaces bat's home with chock-a-block monoculture that the bat isn't used to, and, all things considered, doesn't really like very much at all. Bat flies off and pees in the human's pig farm. Ashes, ashes,[9] we all fall down.

Indeed, even if our goal is addressing climate change on its own—and to hell with the other two problems—we cannot pursue this goal by simply growing chockablock monoculture forests, for these may not last very long. Consider, for example, the longleaf pine.[10] The longleaf pine is a tall, majestic tree that used to dominate the southeastern United States. Before the arrival of the Europeans, it was the dominant species on 60% of the southern forestlands. Unfortunately for the longleaf pine, it had the great misfortune of producing high-quality lumber, an attribute that saw it heavily logged and disappear from much of its ancestral range. This was a range to which longleaf pine was very well adapted. The tree was not tightly packed into a forest, but tended to occupy small groves and, more commonly, spread out over a semi-open savanna. This is no accident. One of the major hazards facing trees in that part of the world is the frequent lightning strikes which often occasion forest fires. This is what kept the landscape semi-open, rather than densely forested: other trees in its midst were continually thinned out by these fires. Longleaf pine, however, had adapted to this problem: it evolved special resistance at the seedling stage involving unusually rapid aboveground growth and a very deep root system. Moreover, the space in between longleaf pines is occupied by low-lying shrubs which are also adapted to the frequent lightning-set fires. The attenuation of longleaf pine changed the structure of the land in several unforeseen ways. Slash and loblolly pine took over the land formerly occupied by longleaf. Different types of taller shrubs moved in. These new pines and shrubs permitted the accumulation of dry leaves and flammable dead branches. The result was that lightning-set fires, which had hitherto been confined close to the ground, in shrubs that were resistant to them, now reached the canopy and became much more widespread and destructive.

The lesson to be learned from this, of course, is that, even if climate change were our sole concern (which, of course, it is not), no project of afforestation

---

[9] Or, if you are British: "A-tishoo, a-tishoo, we all fall down." I have to admit, that version makes a lot more sense to me.

[10] Here, I am indebted to Edward O. Wilson's obviously fond discussion. See his *Half-Earth: Our Planet's Fight for Life*, (New York: Liveright, 2016), pp. 176ff.

can simply afford to plant any old trees willy-nilly, wherever we feel like it. The idea that we can proceed in this way is a parody of any sensible afforestation proposal. Instead, the afforestation must involve trees appropriately adapted to the region reforested: to the soil, the climate, the weather, the other native species, both flora and fauna. If afforestation is to serve any of its three purposes, it will have to be appropriate to the local geosphere, hydrosphere, and biosphere—to what Aldo Leopold called the *land community*. It is also true, however, that the land community in a given region may have become grossly distorted, the results of thousands of generations of human pressure. In the end, I recommend letting historical provenance be our guide. Afforestation should be guided by the ideal—which may or may not be fully attainable in various cases, but can still serve as an ideal—of restoring the planet to the way it used to be in the past. But the past is a long time. To precisely *when* in the past we should focus our afforestation efforts will, of course, vary from place to place, region to region. But what this variation depends on, fundamentally, is one, and really only one, consideration: when, in a given place or region, did humans really start to mess things up?

The Other Side of the Mountain

Almost every morning of my childhood, when I woke up, I would look out of my bedroom window and see *Mynydd Maen*. That is Welsh. A curious language, Welsh. Most languages don't have double *d*s. And if they do, they probably won't be pronounced like the "th" in "the" or the "the" in "tithe." The "y" can be pronounced in different ways: like the "o" in "won," or like the short "i" in "win." Generally, when there are two *y*s in a word, the first is pronounced like the former and the second like the latter. The dipthong "ae" generally is pronounced like the long "i" in "mine." Thus, out of my bedroom window I could see a mountain that was pronounced *Mun-ithe-mine*—with a short "i" in "ithe." This collection of sounds means "Stone Mountain." What does this name—in particular, its assimilation to stone—connote? In his poem "Dowlais Top," Idris Davies, a poet from a couple of valleys over, wrote of a place very similar to my Mynydd Maen and captured the essence of places such as these very nicely:

> What is there here but ragged earth beneath a ragged sky
> A bleak, discolored, broken land where only the stray sheep cry.

Not as bleak or discolored as Dowlais Top, of which Davies wrote. Mynydd Maen was lucky enough to contain no easily accessible coal, which spared her the worst ravages of the Industrial Revolution. Purple with heather, simultaneously wiry and bouncy, in summer, but treeless: as far as I could see from my bedroom window, nary a tree to be found on the entire expanse of the mountain.

If you looked at Mynydd Maen from my bedroom to the east, and then let your eyes run south for about 4 miles, you would see an unnatural mound—unnatural in the sense that it is too symmetrical to have been made by nature—on the southern ridge of the mountain. This is *Twmbarlwm*. Don't get me started on how to pronounce that. It is the remains of an Iron Age hillfort, built by the Silures, a local tribe of Brythonic Celts and, I believe, later co-opted by the Romans as a lookout point. The barren, broken land of Mynydd Maen has everything to do with the existence of Twmbarlwm.

The Last Glacial period had transformed Britain and Ireland into tundra and ice sheet. But with the warming earth, trees began to colonize this period from around 10,000 BC. These new trees were very successful, being circumscribed only by high altitude, severe wind exposure, or waterlogged ground. As 3000 BC approached, Britain and Ireland were carpeted with forests: everywhere trees could grow they had grown. This was age of the *Wildwood*. It wasn't to last long. We humans did, or began to do, what we always do. Evidence of large-scale forest clearance can be found as far back as 2900–3100 BC, courtesy of the newly emerged Neolithic culture. The deforestation increased in scope and rapidity during the Bronze Age (1700–500 BC). But, still, the vast majority of the country still remained Wildwood. But when the Iron Age Celts arrived—including the people who moved into Wales and the Marcher Lands and became known as Silures—the days of the wildwood were numbered. Over a period of several hundred years, the Celts rained down destruction on the Wildwood, remorselessly converting it to coppice, to arable land and to pasture. The Romans then came and conquered the Celts and continued the process of deforestation. By 1000 AD, only 10%–20% of Great Britain and Ireland were covered by forests. The wildwood became packaged into smaller and smaller parcels, separated from each other by grazing land, and with reasonably defined boundaries. The landscape of modern Britain had emerged: farmland, fields, meadows, lined with hedges, and with small, isolated islands of forest. And soon, too small to be called forest anymore, these diminishing islands became simply vestigial woodland.

When I was very young, I used to think trees couldn't grow on Mynydd Maen because it was too high: above the tree line. Life later educated me on what a mountain really is, and Mynydd Maen was barely a mountain, standing not much more than 1500 feet above sea level. But the scales were lifted from my eyes not by increasing knowledge of world geography, but by a trip the Rowlands family took one day, a day when I was still a little boy, to the other side of the mountain. Trees grew on the other side of the mountain! Trees as far as the eye could see, growing glorious and unbound, all over the western slope of Mynydd Maen. Cwmcarn Forest, an early afforestation project. This was my first experience of afforestation, and it seems to have made a lasting impression.

The reason there are no trees on Mynydd Maen—not on my familiar eastern slope at any rate—is not because it is too high, and not because the winds were too harsh. It is because we cut down all these trees and never allowed them to grow back. The job of cutting them down was carried out by our ancestors. The job of never allowing them to grow back is continued by us, via our unwitting agents: where only the stray sheep cry. Sheep happily eat grass, but they really *love* young tree shoots.

Idris Davies was writing of a land disfigured by the nineteenth-century industrialist robber barons. But, in geological time, the 2000 years separating these men of industry from the Silures and Romans is merely the blink of an eye. If, in the focusing of our restorative efforts, our guiding consideration is when humans really started to mess things up, we would have to go back to before the Silures, back to the Neolithic. This will be true for much of Europe. When the Last Glacial period began to wane, beginning around 16,000 years ago, the forests of Europe began their great northward march. The boreal forests of northern Spain and Italy commenced their long journey up through Germany and continued on into Scandinavia. They were followed by the deciduous forests, pushing into the space their boreal cousins had left behind. By 7000 years ago, oak trees had reached southern Scandinavia. These vast European forests remained more or less intact until roughly 6000 years ago, when the late Neolithic farmers and their Bronze Age descendants started to do some serious damage. It is the way forests were in this period, just before the stone and bronze axes started swinging, that should provide the ideal target for our restorative efforts in much of Europe.

In other parts of the developed world, the desired date will depend on when, at least roughly, humans really started messing thing up. In the Americas, for example, perhaps 1607, with the founding of Jamestown, might provide a rough date to work with. In other parts of the New World, dates will vary. In Australia, the beginning of the colonization of New South Wales in 1788 might provide a rough working date. But just across the Tasman Sea in New Zealand, the arrival, in 1300, of Maoris—who slashed and burned large swathes of forest and reduced the moa to extinction—might be a reasonable working date. In the oldest of the Old World—the fertile crescent where civilization, arguably, began—we might have to go back 10,000 years to find a time where the environment had not been seriously compromised by humans. Rough and variable dates, admittedly, but the underlying ideal is always the same: return the land, in so far as we can, to the way it used to be before humans arrived and ruined the neighborhood.

The Secret Song

Davies's poem "Dowlais Top" finishes on a message of hope. In the ecological disaster of this mountaintop, he found, as he put it, "my secret song, my silent psalm of joy."

> For here I found the heart could sing what 'ere the eye could see
> I could sing of the beauty lost and the beauty yet to be.

When I was young I remember regularly traveling up the Eastern Valley by car to see my great aunts, Gladys and Myfanwy—Auntie Glad and Auntie Van—in a place called Nant-y-Glo, which translates as the "Stream of Coal." Up through Pontypool, Abersychan, and past the old Ironworks in Blaenavon, then past the row of miner's cottages of Garn-yr-erw. This was in the 1960s and early 1970s, and the coal industry still held the valley tightly in its grip. The mountainsides were utterly black with coal dust and littered with enormous mounds of coal slag. The slag would sometimes catch alight, with fires burning deep beneath the surface, issuing in smoke columns trailing upward before merging with the always gray sky. These fires could burn for years sometimes. Even the birds woke up coughing in this place. The scene through either side of the car window was reminiscent, more than anything, of hell.

This book may seem grim. I don't intend it as such, and don't think of it as such. Instead, like Davies, I'd like to finish on a hopeful note. Things get better. Even the Eastern Valley is much better today. The scars of the coal and iron industry are still there, of course. But flowers grow there now. The message I'd like you to take away from this book is a hopeful one: *we've got this*! If we want it enough, that is. Look: I'm a philosopher, I admit. In general, we are a rather abstract bunch. What I mean is that specifics are not, in general, our forte. The specific elements of policy and strategy, of the politics and economics, and practicalities of decisions we take: these sorts of things don't come naturally to most of us. Certainly not to me. Back in Chapter 7, I spent a page or two talking about carbon trading and credits—and felt distinctly woozy as a result. *How* things are to get done—that is for politicians, assuming they are still in the business of getting things done. But, if the arguments of this book are correct, then *what* needs to be done—the what rather than the how—is as clear as day. And not a 1960s Eastern Valley day, either. A sunlit Miami day. We need to give up our ancient habit of eating animals. And we need to reforest the land that this makes newly available. We must reverse the historic direction of biomass reallocation that has defined the human tenure on this planet. We must refashion the world into a home for the animals rather than a farm for the animals. We know what we have to do. We just have to make ourselves do it.

I may never see Mynydd Maen again. But I can picture it in my mind, not as it is, but as it was once: an elevated finger of the Wildwood running down from the dark, brooding mountains of the north, past Twmbarlwm to the twinkling *Môr Hafren*, the Severn Sea. And sometimes, like Idris Davies, I let myself succumb to the hope it can be this way again.

Index

For the benefit of digital users, indexed terms that span two pages (e.g., 52–53) may, on occasion, appear on only one of those pages.

CPSIA information can be obtained
at www.ICGtesting.com
Printed in the USA
BVHW051607010223
657047BV00003B/3